University lecturer by day and clandestine ecologist by night, Ben Jacob lives with his family (and a small plantation of native orchids) deep in the West Country.

Praise for *The Orchid Outlaw*

'If this doesn't turn into one of the most talked about nature books of the year, I'll eat my hat. Brilliantly written, urgent and brave'
Lee Schofield, author of *Wild Fell*

'A brave, timely and inspiring example of how to exercise our moral obligation to care for the non-human beings we share the world with. *The Orchid Outlaw* is an empowering tonic to environmental despair'
Leif Bersweden, author of *Where the Wildflowers Grow*

'Ben Jacob turned outlaw by saving orchids slated for destruction, risking prison by tracking down rare species, digging them out in the middle of the night and replanting them in safe places'
Cotswold Life

THE ORCHID OUTLAW

On a Mission to Save
Our Rarest Flowers

BEN JACOB

JOHN MURRAY

First published in Great Britain in 2023 by John Murray (Publishers)

This paperback edition published 2024

1

A CIP catalogue record for this title is available from the British Library

B format ISBN 9781399802284
ebook ISBN 9781399802291

Typeset in Bembo by Palimpsest Book Production Ltd, Falkirk, Stirlingshire

Printed and bound in Great Britain by Clays Ltd, Elcograf S.p.A.

John Murray policy is to use papers that are natural, renewable and
recyclable products and made from wood grown in sustainable forests.
The logging and manufacturing processes are expected to conform
to the environmental regulations of the country of origin.

Carmelite House
50 Victoria Embankment
London EC4Y 0DZ

www.johnmurraypress.co.uk

John Murray Press, part of Hodder & Stoughton Limited
An Hachette UK company

For everyone passionate about saving Earth.

CONTENTS

And the dead tree gives no shelter, the cricket no relief,
And the dry stone no sound of water. Only
There is shadow under this red rock,
(Come in under the shadow of this red rock),
And I will show you something different from either.

T. S. Eliot, The Waste Land

AUTHOR'S NOTE

The names of development sites in what follows are entirely the author's creation. Any similarities between these fictitious names and existing or intended developments are coincidental and no direct correlation should be inferred.

Bee Orchid, from J. Smith and J. Sowerby, *English Botany*, vol. 6 (1797).

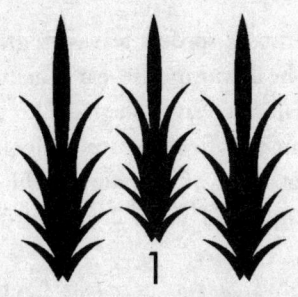

QUIETLY VANISHING

Half-light. Late September. A field. An hour before dawn. Tiny insects bombard my head-torch, looping-the-loop across blacks, greys and silhouettes except where the torch picks out dried stems, withered leaves and brittle seed-head caskets. Few are taller than my knees. No sign of what I am seeking. I continue walking, sweeping the torchlight in slow arcs, searching, conscious that each second is bringing dawn closer.

This is private land. I do not have permission to be here. If caught, I risk imprisonment and fines I cannot afford.

It takes several anxious minutes before the first small, ghostly spiral appears amid the dry grass. Fixing it in the puddle of light, I lower my pack, pull out a blade and push it into the ground. The blade, the kind used to scrape off old wallpaper, sinks in, severing leaves and roots. I pause. The whisperings of the field flow back in. Listen. No heavy footfalls.

Heart pounding, I plunge the blade in again and again, quickly making a circle about 20 centimetres wide, before inserting it for a final time to lever out a ball of earth. The earth resists. Dropping the scraper, I squeeze my fingers in, clutching the planet's heavy, cool, moist-gritty flesh. It becomes a wrestling match: me against the field, me braced against this patch of soil and, more importantly, that one life which has brought me here.

A minute later, the dusty ball is in my hands. I angle the

torch to check the top. It sprouts leaves of grass, a bed of moss, tiny clumps of self-heal and mouse-ear chickweed. In its centre sits a rosette of small flat leaves shaped like green arrowheads. This is why I am in this field at five o'clock in the morning: those leaves and that helix of trembling, white-green, trumpet-shaped flowers with frayed rims. I take a deep sniff. Honey. Rich, sweet honey.

This aromatic orchid, *Spiranthes spiralis*, has the common name Autumn Lady's-tresses. When its flowers appear in the autumn, they pierce the world of light and air like a tiny ornate spear thrust upwards from the underground realm. Over a couple of weeks, that spear becomes a flower spike resembling plaited pale-green strands. Hence their name, Lady's-tresses, the braided hair of Lady Autumn hung with little white-green bells, winding around the stalk, each promising droplets of nectar.

This was one of the first orchids ever recorded in English. William Turner, a sixteenth-century theologian and naturalist, described it in *The Names of Herbes*. Published in 1548, the book was revolutionary in recording the 'commune' (common) names of the plants used by healers alongside the more respectable official ones in Latin and Greek. Turner describes a plant which 'bryngeth furth whyte floures in the ende of harueste, and it is called Lady traces'. They are particularly prevalent, he mentions, at Syon.

Turner's book appeared in the aftermath of Henry VIII's split from the Roman Catholic Church, when Syon was a recently dissolved abbey on the northern bank of the River Thames. I imagine dappled riverside pastures with a ghostly haze of these delicate flowers, their sweet perfume mingling with autumn musk, their incongruously large bumblebee pollinators buzzing between them.

A few centuries later, all traces of Syon's abbey have gone. So too its fields of Autumn Lady's-tresses. Their disappearance was probably gradual, the result of the habitat being ploughed up for farmland and then built on as London sprawled outwards. Much of the abbey's estate was incorporated into that of Syon House, now the residence of the Duke of Northumberland, but no Autumn Lady's-tresses remain. From a sizeable population worth remarking upon to no plants at all – this is typical of the quiet vanishing that has occurred across the country. Today, this once locally common orchid is extinct across large tracts of Britain. It will soon be extinct in the field where I stand too.

Twenty metres to my right, two large banners rise above the black embroidery of a dishevelled hedge and a stark new steel fence hung with signs. 'TRESPASSERS WILL BE PROSECUTED.' A gust sends the banners chinking against their metal poles. 'Coming Soon – The Meadows – Development of Luxury 2, 3 and 4 Bedroom Homes.' The name of this development and that of so many others – Cherrytree Vista, The Elms, West Orchard – are ironic reminders of what they are destroying. The meadows will be gone, cherry trees, elms, orchards and orchids too, native flora and fauna razed.

At the top corner of the field, near the banners and fences, a pair of Portaloos and a command-centre Portakabin have been deposited. Next week, the armoured division will begin their offensive. Diggers will uproot the hedges and engineer this gentle slope into foundations and pipeline trenches. The orchid colony will vanish. A year from now, this earth will be smothered in asphalt, paving slabs, brick, plastic, glass and concrete.

In accordance with English law, the construction company carried out environmental reports and impact statements for this site. Why do they not mention orchids? First, many native British orchids are easily recognisable only for the few weeks each year when they have flowers or buds. Although government guidance stipulates that field surveys 'should' be carried out at the correct time of year and 'where possible' developments 'should avoid affecting plants, fungi or lichens', construction companies have the financial power and prerogative to focus on 'should' and 'where possible', safe in the knowledge that destroying protected plants is not considered unreasonable if that destruction is 'an incidental result of a lawful action'.[1] Constructing an approved housing development is, of course, a lawful action.

Autumn Lady's-tresses did not spend millions of years evolving flowering times to fit in with the recommendations of today's national environmental impact guidelines. Autumn Lady's-tresses flower from late August. When not in bloom, their leaves are inconspicuous, so any survey done in July (as this survey was) is unlikely to find them. What's more, it can take over a decade for this orchid species to produce flowers. No one is sure what makes their little spires suddenly appear and then sometimes disappear for years, but unless field surveys are completed over many years, these rare plants can be missed. No development company waits that long. There could be hundreds here under-ground. Although planning applications include exhaustive surveys of site geology, none consider the unseen life within the soil.

For these reasons the clod in my hands is far more than earth containing a plant that needs rescuing. Until a few years ago, it had never occurred to me that soil is a product of timescales poorly grasped by humans. Innumerable cycles of ice, wind, rain and sun, along with the action of fungi, lichens and plants, cracked, splintered and ground bare rock to form this substance. Millions of years of volcanic ash fell, generations upon gener-ations of organic matter were decomposed through trillions upon trillions of ancient, unseen creatures, each extracting and releasing minerals and chemicals into what would otherwise be a planet-wide mass of sand. Without them our planet would be like the surface of the moon.[2]

It takes 10,000 years to produce one metre depth of soil.[3] At about 12 centimetres in depth, a crude calculation suggests I have just extracted a ball of earth that could have taken around a thousand years to form. As its components were laid down, those particles joined a complex web of relationships connecting millions of organisms. Those relationships and those organisms evolved in and on this land over hundreds of millions – and for some, billions – of years. In fact, the combination of creatures living in this clod evolved only in *this* soil. *This* area. Any notion that soil is soil the world over – or even the county, city or one field over – is simplistic. Orchids taught me that.

The world is gathering substance and colour. A crow rasps three times. Another responds. The birds' dark shapes appear, swimming overhead through the lightening sky towards nearby suburbs.

I check the time: 5:20. About thirty minutes before sunrise. It is already light enough to turn off the torch. Soon my presence will be obvious to anyone who cares to look. I move the orchid in its soil nest to a moist cotton bag, gently tucking the cloth around it. Then I take up the blade again. I want to save as many of these orchids as I can. Each one takes time. Of the three dozen little spirals I can see trembling in the growing light, I only have time to rescue four.

This is why I am here. I have a day job and a family and little opportunity to raise planning objections based on the presence of threatened orchids at The Meadows. Even if I did, the number of housing developments – or new runways, power stations, dams, high-speed train lines, bypasses, retail parks, industrial parks, car parks, roads – that have been halted because of orchid colonies hovers around zero. Even the presence of one of the country's rarest orchids was not enough to stop a site from being destroyed.[4] So, in my own small way I do what I think is right: I take the law into my own hands to save a few plant species from being driven to extinction.

But this is about more than a few plants. This is my attempt to hold together pieces of an intricate tapestry far older than humankind, to salvage part of Britain's natural, cultural and national heritage. It is about taking action to protect a future too often overlooked by the whims of policy-makers; it is about the relationship between humans and orchids, and what that reveals about the land left to us and which we leave for future generations. It is about life, our place on this planet and the duty we all have to protect it. It is about trying to save some of its beauty for my son, who at this hour is in bed sleeping (I hope) like his mum. It is about disobeying outdated laws and backward habits and securing a future for all children, for all lives on and in Earth. This may sound strange, but by the end of this book you may realise it is not: humans are bound to orchids and orchids to us.

I gather the four plants and, as a flamingo-pink dawn sets the sky alight, return to the top of the field and squeeze through a gap between the hedge and the steel fence. Looking back at the field feels like a farewell. I do not know for how many centuries wild orchids have existed here. I suspect that the contents of these two bags will soon be the last of an ancient line.

2

ORCHID MAGIC

Sweat sheened the skin of Wilhelm Micholitz. The sun beat down. Broiling heat rose from the sand. Bleached human bones lay scattered across the burial ground, pale and bright. Among them, orchids grew in majestic profusion. Each plant was covered in blooms; each flower, at least three inches across, was a mix of white, rose and crimson. They were *Dendrobium phalaenopsis* var. *schroederianum*, the Elephant Moth Dendrobe, one of the most sought-after orchids in the world.

It was 1891. There were only a handful of Elephant Moth Dendrobes in Europe and all belonged to Baron Schroeder.[1] Britain's orchid collectors were willing to pay almost any price to own one too.

The thing was, mused Micholitz, trying not to get distracted by the human remains around him, no matter how much orchid-obsessed collectors were prepared to pay for one of those plants, they weren't risking the highest price of all. The first time he had come to this island he had encountered human sacrifice and lost every orchid he collected in a fire onboard his ship.[2] This time, well . . . As the local tribe tore orchids off the bones and deposited them in wicker baskets, Micholitz wondered how long the loyalty he had purchased with looking glasses, beads and brass wire would last. Time was against him. The beach would soon be cut off by the tide, and the warlike tribesmen were growing twitchy about disturbing the resting place of their ancestors. It felt increasingly unlikely that he would be able to secure the plants and his own safety.

Several more bribes and a couple of months later, Micholitz made it back to England. 'With great pleasure,' he wrote, 'I announce to you the fact that I arrived here yesterday with my plants in the finest order, and you will at last get striking novelties from this terrible country.'[3] The terrible country was New Guinea. Micholitz's shipment of *Dendrobium phalaenopsis* was auctioned at the main plant-trading centre of the world at that time, Covent Garden, London, for £2,000 (£800,000 today). When Frederick Sander, one of the most successful importers of tropical orchids to Victorian London, asked the orchid hunter to go back and collect more, Micholitz flatly refused. 'I value [my skull] far too much to risk it,' he replied.[4]

The Victorian fashion for tropical orchids was, in part, sparked by the immensely wealthy 6th Duke of Devonshire, William Cavendish. After encountering a tropical orchid for the first time in 1833, he became smitten, and invested millions of pounds (in today's money) in buying exotic orchids and building glasshouses big enough to ride carriages through to house them.[5] The duke's passion was contagious, and when Baron Schroeder joined in, the fashion snowballed and became an obsession of the wealthiest section of society. This fad was variously known as Orchidelirium, Orchid Mania and Orchid Fever.

At the peak of Orchid Fever, the greatest botanical minds estimated that an astonishing 6,000 orchid species existed.[6] Today's best guess puts the number at around 30,000, the vast majority of them tropical. This compares to around 12,000 species of grass and 6,399 of mammal. It is thanks to the ubiquitous presence of tropical orchids in supermarkets and florists, TV and film sets, decorations on books, stationery, wallpaper and textiles, that the word 'orchid' tends to conjure up in people's minds *these* descendants of that Victorian craze. Until ten years ago, that applied to me too.

All orchids have shared traits, whether growing in temperate or tropical zones, but there are huge differences when it comes to size, scent, flowers and preferred habitat. The largest species, the Tiger Orchid, *Grammatophyllum speciosum*, is as big as a Volkswagen Beetle; the smallest, *Campylocentrum insulare*, has flowers only 0.5 millimetres across and was initially mistaken for a fungus.[7] Scents range from decomposing meat to vanilla, chocolate, citrus and honey. The flowers of the little red-haired *Bulbophyllum barbigerum* flip up and down in the faintest breeze like a crowd of punk flies giving a Mexican wave, while the twin metre-long burgundy petals of *Paphiopedilum sanderianum* cascade like regal sashes to provide a convenient highway for ants in Borneo to reach the flower's nectar. In habitats as varied as humid equatorial jungles and mountainsides buried in snow for months at a time, powered only by air, water, sun and wind, orchids have colonised every continent on the planet (except, today, Antarctica). You might consider all this a success story, but another trait they all share is that, worldwide, orchids are dying.[8]

Today's tally places fifty-one native orchid species in the British Isles.[9] Discussion is ongoing as to whether several more species are recent invaders that have spread due to a warming climate. Some of these represent the westernmost outpost of species which range north to the Arctic circle, east to Siberia, the Himalayas and Japan, or south to the Atlas Mountains and parts of the Middle East. A few are also found in North America. Every one is in decline. Some are found in no more than a handful of sites around the country. One species exists in Britain as a lone wild plant. Another might already be extinct. Over the past century and a half some of Britain's orchid species have been thought extinct several times, only to be rediscovered, miraculously, years later.

Only among concerned scientists, conservationists and enthusiasts does this disappearance receive attention. Ten years ago I too didn't care about it. It was only thanks to a series of unexpected events – involving ancient Chinese writings, Asian jungles,

English meadows, the birth of modern science, industrialisation, commercial farming, the British Empire, apothecaries, legislation, climate change, soil biology, a baby boy, a mugging and a broken back (not necessarily in that order) – that I became aware of how wrong I had been. It's been quite a journey of discovery, one that began in Caracas, the capital of Venezuela, when a hand descended heavily on my shoulder . . .

I was returning from forty-eight hours spent photographing wild orchids among hummingbirds and strange, fleshy-leaved, high-altitude plants on a mountain ridge that looked out over slopes of jungle canopy to the aquamarine Caribbean. I had hiked back down to the city, pack and tent on my back, and was making my way through a thronged street, past sellers pushing trolleys piled high with mobile-phone cases and white plastic carts full of ice cream. The noise was deafening: ringing bells and cries competing with the electronic beat of *reggaetón* blasting from nearby shops. Open-air cafés offered olfactory hints of *arepas* and *chicharronadas*, which all mingled with dust, heat, flies and the fumes of bleating, grunting, shunting traffic, like a slow procession celebrating the cheapest petrol on the planet. After time on the mountain away from people, all that banging, clanging, grumbling, hooting, shouting and beating had clogged my early-warning systems.

I snapped my head around. Fingers were gripping my shoulder. They emerged from the dark-blue sleeve of a tunic criss-crossed with belts and holsters. Embroidered in white above the breast pocket were four words: *Policía Metropolitana de Caracas*.

Grinning, the policeman demanded, 'Passport?' He looked as if he had just won the gringo lottery.

This was not about my passport. With the Metropolitan Police of Caracas it never was. At that time the city had the highest per capita murder rate in the world. Statistically, the chances of getting killed were higher there than in Iraq, where a war was raging, and I had just learned that Venezuela's police were responsible for a significant percentage of the city's murders. (Shortly after I was grabbed, the city's entire police force was disbanded – on full benefits – amid reports that they were responsible for a fifth of all local crime.)

Propelled by those strange confluences of events that sweep us through life, I had not actually travelled to Venezuela to become a murder statistic. Ostensibly, I was there to teach English – largely, as it turned out, to moneyed professionals (and the offspring of moneyed professionals) desperate to emigrate to countries where the police were less overtly corrupt. Really, though, I was there for the orchids.

Tropical orchids had enchanted me since my first encounter with one in an English garden centre. I was nine. Admittedly, compared to Venus flytraps, my other obsession, that orchid was more conventionally attractive, but it was also more . . . simply more . . . than any flower I had seen. It might have been teleported from a prehistoric jungle within earshot of a bellowing Tyrannosaurus rex. Its weird and beautiful flower possessed an exuberant prehuman glory. It was starlike, the petals resembling sultry lips sculpted out of wax and then tinted with hues distilled from ruby, cinnabar, coral, specks of opal, washes of topaz.

My first orchid sat on the landing for years, happily growing long, dark-green leaves, but after those first flowers faded, it took a long time, and much more knowledge, before it blossomed again. That orchid was a *Cymbidium*, a native of China, bred to satisfy a continuing worldwide demand for tropical orchids worth hundreds of millions of pounds a year.[10] At that age, I had no idea about GDP, exports, international trade. All I knew was that I liked those flowers. A lot.

So began a lifelong fascination. I was interested in many aspects of the natural world – animals, birds, insects – which my parents encouraged with gifts of David Attenborough books and Saturdays spent birdwatching. Science was part of the school curriculum, and there I dutifully learned a few things about botany – photosynthesis, germination, chlorophyll – but my personal interest in nature never married with formally studying it. That must be why the notion of a career in ecology never occurred to me. Instead, I focused on literature. This didn't lead to a career in botany or conservation, but neither did it mean I entirely abandoned orchids. Rather, studying how stories worked helped me see orchids as more than, in dull scientific terms, perennial herbs of the family Orchidaceae – terrestrial, litho-phytic or epiphytic, sympodial or monopodial vascular flowering monocotyledons. I came to see them as symbols, reflections, living objects which, for centuries, humans in cultures worldwide have endowed with special significance.

I learned that, in ancient China, some of the first recorded descriptions of orchids are offered in the *Shen Nong Ben Cao Jing* ('The Divine Farmer's Materia Medica Classic'), a text with origins lost in the half-mythical mists of Chinese history, when divine beings descended from Heaven. They are presented as medicinal herbs with applications ranging from prolonging life to stopping bleeding, quenching 'vexatious thirst', promoting lactation, removing small worms, brightening the eyes and coun-tering the effects of ageing.[11]

Centuries later, the wandering scholar known to the West as Confucius consolidated the importance of orchids in the Orient when, so the tale goes, he stumbled across a woodland grove full of orchids in bloom filling the air with sweet perfume (the Chinese for orchid – *lan* – also means 'fragrant'). For Confucius, those plants created beauty whether there was an audience or not.[12] They were emblems of determination, humility and true nobility. As Confucianism spread, those desirable traits trans-formed *lan* into more than medicine: the orchid became the aspirational model for millions, the subject of poetry and art, and one of the four noble plants of Oriental culture.

In Central America the Totonacs and Aztecs used vanilla – the dried seed pod of the Vanilla Orchid (*Vanilla planifolia*) – to flavour their chocolate. Spanish colonisers learned its secrets from the Aztecs, other Europeans from the Spanish, and even today its labour-intensive production means vanilla remains one of the world's most expensive spices.[13]

The European love affair with tropical orchids soon extended far beyond vanilla. Between the early and mid-nineteenth century, European colonization, trade, military and missionary networks opened up new areas of the world. Commodities – cotton, wool, sugar, tea, rice, silk – flowed back in larger quantities than ever. Alongside these came rarer goods: hardwood, ivory, ostrich and bird-of-paradise feathers; the carapaces of dead beetles and butterflies to adorn the hats of wealthy ladies; tortoise shells, crocodile skins and beautiful exotic plants. These luxuries fed a growing demand for the 'exotic'. It was in the early years of this influx that William Cavendish encountered the tiger-striped bloom of an *Oncidium papilio* and became enchanted. Soon after that, owning tropical orchids became a vogue, an industry, a competition among the wealthy.[14]

Commercial plant nurseries sprang up, but long before the discovery of cloning and micropropagation techniques, nurseries could only source tropical orchids in the wild. Plant hunters such as Wilhelm Micholitz trekked through malarial swamps and jungle, up treacherous mountains, and through rapids. They were prepared to endure floods, disease, dangerous animals, shipwrecks and hostile indigenous tribes in order to locate, extract and transport exotic plants – in particular orchids – to Europe, and perhaps most of all to Britain.

The tales of Micholitz and other plant hunters, like Benedict Roezl, a Bohemian with an iron hook in place of his left hand, who spent years scouring South and Central America extracting tons of orchids (a statue of him, without his hook, can be found in his native Prague),[15] and Joseph Dalton Hooker, who explored the Kingdom of Sikkim in the Himalayan foothills (and who was imprisoned there by the Dewan of Sikkim for straying into prohibited territory, and later became director of the Royal Botanic Gardens, Kew), filled me with awe. Their names were immortalised in the rare orchid species – *Phalaenopsis micholitzii*,

Miltoniopsis roezlii, Cymbidium hookerianum – that they collected and sent to Covent Garden. Orchids like these sold for astronomical sums. Consider, for example, the first commercially available *Cypripedium spicerianum* which was sold in the late 1870s for £250 (approximately £100,000 today).[16] In 1881, Frederick Sander noted that the 1,000 *Cattleya labiata* he had imported had the potential to fetch £10,000 (more than £4 million today).[17]

By the end of the nineteenth century, tropical orchids were valuable living treasures. Part of a new-found commodification of the natural world, their appeal combined fantasies of far-off lands and elements of exploration, conquest, adventure and discovery. Thousands of trees were felled to collect the orchids growing on them; thousands more plants died on sea-crossings, and many of those that survived did not last long in the care of owners with far more money than botanical knowledge.[18] In China, orchids had been medicines, aesthetic subjects and models of desirable nobility. In Europe, they were objects of luxury and obsession; makers of fortunes, fuel to an industry and a fever for collection.

I started to wonder whether I might follow in the hunters' footsteps to find exotic tropical orchids in their native habitats. I was not a botanist, ecologist or scientist. I wasn't engaged in research. I simply wanted to see them where they belonged: I wanted to see what those orchid hunters had seen.

Unfortunately (or fortunately), Orchid Mania had faded with the First World War, and my international travel would have to be self-funded. I managed this by washing dishes in restaurants, working night shifts on the production line of a margarine factory, packing tractor parts, tidying up a building site and picking parsnips. My earnings allowed me to buy cheap flights to far-flung places. So, armed with rucksack, camera, traveller's cheques and guidebook, I set off to find orchids.

The Thailand, Malaysia, Cambodia, Sumatra and Java I experienced were utterly different from those encountered by Micholitz, Hooker and many others. I took in crowded bus stations, neon lights, skyrise office blocks. I didn't see tigers and didn't experience a shipwreck. Hours on planes, local buses and tuk-tuks replaced months on ships.

On the upside, I had hiking boots, Internet cafés, malaria pills and vaccines against yellow fever and Japanese encephalitis. On the downside, I lacked a rich employer to send me additional funds, and didn't have a team of porters paid in copper wire to carry my bag and cook all my meals. But when I gazed out over cacophonous jungle canopies, a ghostly vapour luminous above spectral trees in the light of the rising sun, it felt as if I glimpsed something those plant hunters had seen too.

Like the orchid hunters, I soon learned that, save for occasional surges of rattling cicadas, in the steamy heat of the day, jungles are mute. Trekking through them is like wading along the floor of a green-and-brown ocean. Life operates at a different pace and scale. It is a slow botanical stampede for the sky: bright splurges of tiny fungi, flashes and flickers of iridescent wings, trickles of song from unseen birds; legions of insects, arachnids and annelids, seething, writhing, scuttling, whirring.

I learned which jungle microclimates orchids favoured and I tuned in to the subtle shifts in light and humidity of areas most likely to harbour them. If the season was right, I discovered orchids flowering, usually perching on mossy branches, their flowers shining like gold, embers, gems, in the rich green darkness.

Many evaded me and my camera. They grew out of sight in the jungle canopy, thriving in occasional downpours and tropical light. Unable to climb up to them, I had to make do with what I could find from the ground. Then I found out that many tropical orchids grow on mountain tops, where conditions are perfect for multiple species. There, they cling to stunted trees or rise from carpets of moss, basking in the moisture of passing clouds, bright days and cold nights. Most importantly (as far as I was concerned), they were within sight of an orchid explorer like me. Orchids and mountains – they were what led me to Caracas.

★

At the time, Caracas was home to about three million people, with another million or more packed into surrounding shanty towns. The city sprawled across a wide valley 900 metres above sea level. To its immediate north, a jungle-clad ridge rose over 2,000 metres then plunged to a strip of Caribbean coast. Officially this ridge is the Waraira Repano National Park. Locals call it 'El Ávila'. It dwarfs the city's steel-and-glass skyscrapers. *Barrios*, the vast jumbles of often gang-controlled shanty towns, lap at some of El Ávila's lower slopes, but elsewhere across its peaks and broad, less accessible slopes, that ridge offers a big slab of wild mountain. Every window of my thirteenth-storey apartment overlooked the mountain's mist-shrouded slopes. I knew orchids were waiting there.

My first visit to El Ávila turned out to be a twenty-minute walk along quiet suburban streets past mango trees and tall barbed-wire-topped walls. There, at the foot of El Ávila, a faded green-and-yellow sign and overflowing litter bin marked the entrance to the park. I wandered in. No gate, no ticket office, no ranger station. Just a couple of curious bright-green-backed, electric-blue-headed jays watching me. I passed beneath the soaring trunks of trees. High above, a dense canopy obscured the sun. Through twilight I followed a muddy track. It forked and forked again and wound along the mountain flank. Trees, palms, ferns, vines, bamboo and walls of plants I did not know grew in rich profusion. Motes of flying insects shone in stray rays of light like tiny swirling dancers.

I followed the path upwards. Sporadic breaks in the trees offered glimpses of skyscrapers, grey rivers of winding highways and twinkling cars. I climbed higher. The tracks were deserted, save for an armadillo, which bolted, apparently astonished that a human had bothered to take that route. The further I went, the more overgrown the paths became, until most petered out, reclaimed by jungle. With evening near, I decided to turn back. As I hurriedly retraced my steps through thickening gloom, I had my first encounter with a wild Venezuelan orchid. Sitting on an overhead branch, a clump of yellow-orange starry flowers peered at me like jovial faces. Probably a *Miltonia* of some kind. It was too far away to be sure and it was too dark to take a decent photo, but I didn't mind. There seemed to be plenty of paths there and my contract to teach English was for two years.

On my way back to my apartment, the city felt strangely unreal. Artificial lights, concrete and cars, garrulous diners at plastic tables – it all stood in stark contrast to the world I had emerged from. Passers-by glanced at me – who is this scruffy gringo drifting through our affluent neighbourhood? Even I started to wonder, as I walked through the city, a foreigner in more ways than one, which world I – or any of us – really belonged to.

Over subsequent months, in between days spent in a tower teaching grammar, I discovered El Ávila's orchids. The main route to the mountain ridge was a six-hour trek through half a dozen vegetation zones (it could be done in three, but what was the point in racing when there were so many plants to see?). Along the way, the tall palms of the humid lower slopes gave way to cloud forest, where trees were draped in mosses, ferns and vines and clouds raced past like grey shadows. Above that were plateaus of strange, thick-stemmed, hairy-leafed plants. Above them the mountain ridge, strewn with boulders and stunted trees bearded with lichen. In some places the ridge was no more than the width of a path, a sheer drop on each side plummeting hundreds of metres. To the south lay Caracas; to the north, the wide blue Caribbean.

Whenever my teaching allowed, I struggled up to the ridge with water, food and a tent, and set up camp. Mostly I encountered species from genera like *Epidendrum*, *Schomburgkia*, *Oncidium* and *Cycnoches*, with starry scatterings or clusters of intricate flowers, growing as I imagined they had done for centuries, buzzed by emerald-green-and-blue hummingbirds. I photographed them and took notes of what I found before the sweltering day became an almost freezing night. Here was proof, if proof was needed, of the misguided beliefs of all those Victorian orchid collectors who kept their tropical orchids in stiflingly hot, dark, humid greenhouses and consequently caused their very expensive plants to quickly expire: in sight of the Caribbean, I was shivering on a mountain surrounded by orchids, too cold to sleep.

Caracas was a dangerous place. In hindsight, wandering around that mountain by myself was idiotic. Setting aside the dangers of any hike (sprains, falls, breaks), it was a vast uninhabited area neighbouring a city with an astronomical murder rate. I reasoned the chances of being mugged or murdered on the mountain were slim because no gun-wielding bandit would bother scaling a mountain to rob someone when the number of potential victims in the city was far higher. Thankfully, experience proved me correct: the few people I encountered on the mountain trails were other hikers.

What I hadn't taken into account was the walk from the jungle to the apartment. Perhaps my own variant of Orchid Fever had impaired my logic. After all, the same rationale that rendered the mountain relatively safe suggested the urban walk was not – as I discovered one hot day after a sleepless night on the ridge and a long trek down the mountain. That was the day the Hand descended onto my shoulder.

A black police truck was parked nearby. It had a sliding side door. The door was open. Four uniformed officers, pistols holstered, lounged inside. One was female, two were younger guys, one older. The older guy was in charge. Their faces lit up as the Hand shoved me towards them. One stepped out to meet me in a way which was less about hospitality than perusing the catch of the day.

Instinctively capitulating to the will of armed men in legit-imate positions of authority, but suspicious of what might unfold, I tried to keep my distance from the truck's door. The officers rattled off commands and questions in Spanish too fast for me to grasp. The gist was they wanted me in the truck. An old lady walking past remonstrated. They told her to mind her own business. She responded with angry words. When the police threatened her with her own interrogation in the truck, she sadly shook her head and drifted away. I was on my own.

They wanted my passport. I had a photocopy in my bag. I got it out, handed it over. They said it wasn't valid without a copy of the immigration stamp. I knew that wasn't the case and tried to say so, but they were not interested. I tried to bluff my way, saying I worked for the British Embassy. The female officer suggested they let me go. The older guy disagreed. He wanted to search my bag. One of his cronies was already tugging at it. Reluctantly, I let him take it. I wondered if that was the moment to bolt for safety, but at the same time I didn't want to give them an excuse to shoot. I also didn't want to abandon my orchid notebook or my camera loaded with precious photos. They took the bag inside the truck, threw it on the table and surrounded it. I couldn't see what they were doing. I would have to go inside.

With heavy heart, I stepped in.

They were gutting my bag. One of them found the crumbly remains of a paracetamol tablet which had probably been lying at the bottom of a side pocket for years. He handed it to the older guy, who studied it, sniffed it. Looked at me.

'What's this?'

Gravely, the old police officer shook his head as if he had just captured a drug kingpin and I was facing several lifetimes behind bars. In the meantime, the others had extracted my notebook, tent, camera and empty food boxes. One turned the camera on and, like a moron who had never seen such an invention, planted his thumb right in the middle of the lens.

'*Cuidado!*' I exclaimed. 'Careful!'

'What's in there?' barked the older guy.

'It's a camera.' Did he seriously think that pocket-sized camera was stuffed with cocaine?

Instantly the mood soured.

The older guy jabbed a finger at a partitioned area at the back of the truck.

'We have to search you,' he said.

Even I wasn't stupid enough to enter that concealed area.

'Have the camera,' I pleaded. 'Please give me back my book.' I tried again in case my Spanish was wrong, with clear, simple gestures.

Camera you. Book me. I go. Yes?

The female officer looked from me to her colleagues. The guys were in charge. The Hand grabbed my arm and yanked me towards the back of the truck. I twisted free. The officer's own momentum made him stumble. He whirled back with hurt pride. The small space filled with noise. Caught between doing what I was told, my notebook and the bright street where someone might – might – come to my aid, I froze. In that moment, the officers descended on me and it was too late to escape. I remember fear bearing down on me like an intense, ominous shadow. Then darkness.

Some time later I rose up through that dark until I reached a subdued artificial glow. A curtained cubicle. Tubes. Stainless steel. Machines with coloured lights, dials and numbers. The shuffling of people along a corridor. Was it the afterlife? A dream?

I am not sure how I got there. Details of my hospital stay are hazy. Was it one night or two? Which colleagues came to my bedside? Did I ever thank them? Did I eat? Drink? I dimly recall a doctor with a white beard conducting tests with electrodes on my head and showing me charts with wavy lines traced on squared paper. He said I might have brain damage.

I had lost everything. When I returned to my apartment, the door was open, the rooms empty. My passport, laptop, money, even most of my clothes were gone. The loss of most 'things' didn't bother me – they could be replaced – but the loss of items with absolutely no monetary value filled me with hurt:

the contents of that notebook and the photos on the camera and computer. The only images left were those I had emailed to people, and they were a fraction of what I had taken. I had lost all that, all because of orchids. Damned tropical orchids.

In a rare moment of clarity, I realised it was time to give my orchid-influenced globetrotting a rest. It was time to return home while I still could. Life would be much safer in England. Or so I thought.

3

LESSONS OF A BEE

Peering out of a porthole at 30,000 feet, I saw England spread far below like a patchwork quilt. I had fled Venezuela as soon as the British Embassy issued a replacement passport and was returning home to recuperate, refinance, re-acclimatise to a life without the dangers of seeking tropical orchids. England was welcoming, familiar, comforting. Small towns, Norman churches, bobbies on the beat. Fields bordered by dark hedges draped across undulating land straight from the pages of Jane Austen and Agatha Christie and the canvases of Thomas Gainsborough and John Constable. Ah, England. Mild, lush, fertile, its history, art and culture rooted in that damp temperateness.

With only an empty rucksack to my name, I retreated to my parents' house in Devon. I secured a short-term job teaching English and, lacking jungles to explore and orchids to find, decided to take up running.

One damp morning I ran through a park and down some broad marble steps. I underestimated the marble's slipperiness. My shoe, and everything attached, skated into thin air. As neatly as if an invisible giant had scooped me up and dumped me, I flipped, fell and slid down the flight of twenty steps, each impact 'cushioned' by my back.

Seconds later, I was blinking at the sky. Something serious

had occurred. What felt like a searing hot dagger impaled my spine and lightning ripped through my ribs. Cautiously, I wiggled fingers and toes. I could feel them. That was some relief, but it didn't stop the excruciating pain.

It was early morning. The park was empty. I had dropped my phone during the fall. I couldn't see it. I tried to stand. I'm over six foot tall, with shoulders to match, but the entire world was swimming with pain-induced tears. I spotted my phone, then, somehow, very, very, slowly reached it and, for the first time in my life, dialled 999.

An ambulance trip to the nearest A&E department confirmed a cracked vertebra. If it had been worse, I could have been paralysed. If the impact had been about 20 centimetres higher, on my skull, I could have died.

Over the years I had broken several bones, but the cracked vertebra established itself as number one in the Painful Fracture League. Forget sneezing or coughing: each was like smashing the bone on marble again. Movement had to be very slow so as not to wake the sadistic ogre who had taken up residence in my spine. I would have to take life easy for a while.

After a couple of days, restlessness set in. For sanity's sake, I began shuffling around the flatter parts of my parents' garden. They are keen gardeners and it is a well-curated place. During the years I had been away, my father had let part of the lawn grow into a mini-meadow. The long grass rippled in the summer breeze; sooty-winged gatekeeper butterflies oscillated between buttercups and ox-eye daisies. Among them, the quivering of some small milky-pink pennants caught my attention. Curious, I lowered myself to one knee, sucked up the hurt and leaned in.

The three little pink flags were pointed petals (I'll call them petals for now and get into orchid flower morphology later), arrayed a bit like the horns of a jester's hat (one pointing up, one each to the side), around a fuzzy brown bumblebee-sized oval. The markings on that brown velveteen oval reinforced its comic appearance: pale yellow and curving upwards, it looked like a grin. An odd long, thin lime-green protuberance overhung the oval. From it hung two small sunny baubles, like a pair of botanical dice hanging from a rear-view mirror. I was astounded. I recognised it. I was astounded *because* I recognised it. There, in my parents' overgrown lawn, was an orchid.

Like the tropical orchids I had hunted across the world, that little flower's lower petal was bigger than the rest – it formed the velvety oval at the centre of the pink triangular flags. Above it, either side of the long, thin, green overhanging structure, two more petals emerged, so slim a casual glance might overlook them. They resembled insect antennae.

The little flower was completely captivating, unlike anything I expected to find growing in England, let alone in my parents' garden. I knew of several tropical species that had developed flowers to resemble insects, and I knew Britain's Bee Orchid employed the same modus operandi, but I was stunned to see it there, stunned not only by its elaborate appearance, but because it was the first time in over two decades of interest in orchids that I had seen one of England's own species.

As my eyes tuned in, I realised there were more little pink flags among the grass stems, each with a bijou bee-like nugget at their centres. Many more: four on that first stem and up to six on half a dozen other long, upright stems punctuated the mini-meadow. Why were they there? How had they appeared? Where could I find more?

Yes, Britain has a few resident orchids and I knew enough to recognise that one, but if you had asked me to name another wild British orchid, I wouldn't have had a clue. How many wild orchid species were there in England? Certainly no more than a handful; maybe only the Bee. I didn't know. I had simply never given them much consideration. After all, what is interesting about something growing in a field down the road compared to the starry yellow orbs of a dozen *Epidendrum elongatum* swaying in a mountain-top breeze within sight of the

Caribbean as shimmering sapphire hummingbirds whirr past? For the entire time orchids had guided my life, I had relegated European and British species to a second or even third league. They were uninteresting. Common. Lesser. And yet, that day, crouching in a corner of the garden and wondering if I would be able to stand up again unaided, I realised my mistake. Those little Bee flowers were exquisite, bizarre, complex.

Euphoria stirred. Life in England need not be life without orchids.

Confined to sitting, lying and taking short, slow walks, I decided an interesting diversion would be to acquaint myself with the Bee's native brethren. I ordered the second edition of Anne and Simon Harrap's *Orchids of Britain and Ireland: A Field and Site Guide*.

Its 400-plus pages were a revelation. They were crammed with intriguing photographs of flowers shaped like little people, insects, boiled eggs with a spoonful taken out of the top, graceful winged birds, tassels, columns and pyramids; pillars of flowers ranging from moon-white, through sun-yellows and entire palettes of pinks, magentas, violets, purples and scarlets. Some plants were smaller than my thumb; others had flower spikes a metre tall. Some (apparently) smelled of goat or honey or cloves; all were accompanied with names far more relatable than the Latinate binomial forms used to identify tens of thousands of tropical orchids (*Brassia verrucosa, Cattleya labiata, Paphiopedilum sanderianum*). Instead there were simple names like Lady Orchid, Military Orchid, Coralroot Orchid, Sword-leaved Helleborine, Fly Orchid, Ghost Orchid, Burnt Orchid. For the first time, I took in the sheer number, beauty and variety of Britain's fifty-six species of wild orchids (as calculated by that edition of the Harraps' book). *Fifty-six!*

The Bee's Latin name, *Ophrys apifera*, was deciphered there: *ophrys* from the Greek for 'eyebrow', which may refer to the furry appearance of many flowers in the genus; *apifera* from the Latin *apis* for 'bee'. The Harraps also explained that the first

British record of a Bee Orchid appeared in 1597 in a book by John Gerard, *The Herball, or Generall Historie of Plantes*. I had never heard of Gerard. This puzzled me. Who was he? What had he said about the Bee? I set the Harraps aside, found Gerard's *Herball* online and, in the space of minutes, lost myself in yet another whole new world.

I didn't fully appreciate it at the time, but that Bee Orchid had led me back to a sixteenth-century European botanical renaissance, a time when long-established sources of plant lore by Ancient Greek and Roman writers were suddenly being questioned. Printing and the use of woodcuts for illustrations combined with the many new species observed and collected by travellers in the recently discovered Americas had reinvigorated the study of botany. Illustrated books on plants suddenly abounded. *The Herball* was one.[1]

Looking like a book of spells spliced onto a forgotten manuscript by Shakespeare, it opened with a copperplate image of scrolls, cherubs, pillars and roses, followed by letters and poems extolling the expertise of John Gerard, barber-surgeon and herbalist extraordinaire. Despite these glowing recommendations, by today's standards not all *The Herball*'s entries are reliable: it is a text in which myths and facts overlap. Gerard classes coral as a moss and claims that when drunk in wine it 'provoketh sleepe'.[2] Another entry recounts magical waters that turn any object to stone, while another describes the 'Goose' or 'Barnakle' Tree, 'founde in the northern partes of Scotland', which gives birth to geese.[3]

Anticipating a fascinating explanation, I searched the 1,300 ornate pages to see what Gerard said about the Bee Orchid, *Ophrys apifera*, except . . . no entry for *Ophrys* could be found. Only by cross-referencing the old woodcuts in *The Herball* with the Harraps' photographs could I locate Gerard's chapters on native orchids.[4]

Although the woodcuts were small and crude, the similarities were clear, but Gerard's names – Maimed Satyrion, Humble

Bee Orchis, Greatest Goates Stones, Triple Ladie Traces, which he described for the benefit of other apothecaries as growing 'in the fields adjoining to the pound or pinfolde without the gate at the village called Highgate nere London' – had no equivalent in the Harraps' guide, just as the Latinate names Gerard used (*Orchis Myodes Minor* for the Small Yellow Satyrion and *Orchis Strateumatica Minor* for Souldiers Cullions) belonged to the long-lost era in which Highgate was a village (since Victorian times it has been an expensive suburb, swallowed by the spreading capital).

Eventually, I located the first ever description in English of a Bee Orchid. It wasn't *Ophrys apifera*. It was, at least as far as I could tell, the Great Humble Bee Orchis (*Testiculus Vulpinus Major*) or maybe it was the Waspe Orchis (*Orchis Melittias*). Gerard included both in a class of plants (which Linnaeus's system later eradicated) called 'Foxestones'. 'There be divers kindes of Foxestones . . .' wrote Gerard, 'some have flowers, wherein is to be seene the shape of sundrie sorts of living creatures; some the shape and proportion of flies, in other gnats; some humble bees, others like unto honie bees; some like butterflies; and others like waspes that be dead.'[5]

Did a British orchid with flowers like 'waspes that be dead' exist? If it did, I wanted to find one. I checked the Harraps' book. No Wasp Orchid. Was it the Fly Orchid, known today as *Ophrys insectifera*, with sleek futuristic blooms, but not much like a wasp 'that be dead'? Was the Wasp Orchid a fanciful creation like the 'Barnakle Tree'? There was also the question of Gerard's Humble Bee Orchis, which, after leafing through the Harraps' book, I thought might be today's Green-winged Orchid (*Anacamptis morio*), but I couldn't be sure. Were they the same species? Why, over time, had it lost its insect associations? Or had Gerard's Humble Bee gone extinct? Whatever you called them, it was clear from those two books, published four centuries apart, that Britain was home to quite a variety of orchids and a rich seam of orchid-related medical and botanical history.

In Gerard's day, orchids were common. He refers to them growing 'profusely' in old chalk pits, 'moist and fertill meadowes', 'pastures and fields that seldome or never are dunged or manured', 'moist and waterish woodes', 'plentifully in sundry places, as in the fielde by Islington neere London, where there is a bouling place under a fewe old shrubby okes', 'upon barren chalkie hils and heathie grounds, upon the hils adjoining to a village in Kent named Greene-hithe' and 'upon the declining of the hill at the North ende of Hamsteed heath'.[6]

Another notable detail was that Gerard's book was not aimed at plant-spotters. It was for 'herbalists', healers who used medicines derived from plants. Gerard was one. For him and all medics of the age, orchids had a use. That use was neither the Orient's floral embodiments of modest determination nor Empire-infused fantasies of exotic places. In Gerard's day, interest in orchids was all about sex.

'Our ages useth all the kindes of stones to stirre up venerie,' Gerard observed.[7] 'Stones' (as in 'foxe stones', 'dog stones', 'fooles stones' and 'goates stones') was the name for many orchids in his day. It was also the name for the smallish oval underground tubers from which most British orchids sprout every year, a bit like the bulbs from which tulips and daffodils grow *and* the Elizabethan term for 'testicle'. Gerard's chapters on orchids abound with references to 'Cullions', 'Ballocks' and 'Cods' – all Elizabethan terms for the same thing. He allocates Ladie Traces (today's Autumn Lady's-tresses) the 'scientific' name *Testiculus odoratus*, literally 'sweet-smelling testicles', and assigns others the label *Orchis* (Butter-flie Orchis, White-Handed Orchis, Spotted-Birdes Orchis) – *orchis* being the Greek for 'testicle'.

Gerard also describes several 'Satyrions' (Small Yellow Satyrion, Bird Satyrion, Maimed Satyrion), referring to the highly sexed, half-beast satyrs of Greek mythology, and he gives the Butterfly Orchid the Latin name *Hermaphroditica*, denoting Hermaphroditus, the androgynous child of Hermes and Aphrodite (goddess of love, beauty, pleasure, passion and procreation), all three being connected down the centuries to the erotic, sensuality and fertility.

The woodcuts in *The Herball* dwell at least as much on the orchids' tubers as on their flowers. The tubers were what interested the herbalists, apothecaries (professionals who created medicines, often, but not only, from plants) and their customers; the flowers were only convenient for identifying the useful underground parts, which, according to *The Herball's* illustrations, almost universally consisted of a generous pair of orbicular tubers.[8] Tropical orchids don't have underground tubers. Many tropical species wrap fleshy roots around the branches of trees and store nutrients in swollen stems called *pseudobulbs*. So, while the whole concept of orchids growing from underground tubers was new to me, the fact that they were once aphrodisiacs was mind-blowing.

A bit of detective work led me to where the idea of orchids being connected to sex came from: the 'Doctrine of Signatures'. It is a concept that pops up in various cultures in which the medical use of a plant is based on its appearance. For example, a mushroom cut in half resembles the human ear, so mushrooms were used to cure earache. The leaves of the cowslip primrose (*Primula veris*) look a bit like lungs, so medicines were made from them to cure breathing ailments. Flowers of the herb eyebright (*Euphrasia*), with a colourful centre in their circular white flowers, look somewhat like eyes, so infusions of these were used to treat eye complaints. The growing roots of saxifrage or rockfoil (*Saxifraga*) can break apart rocks, so it was thought to relieve kidney stones. It's an ancient way of interpreting the world, but many of the same cures can be found in herbal outlets today. It also explains why Europe's orchids were associated with sex.[9]

The orchid–sex connection was first expressed in writing by an Ancient Greek botanist born in the fourth century BCE, Theophrastus of Eresus.[10] Theophrastus also coined the term *orchis* after the testicle-shaped tubers of many terrestrial orchid species (versions of that ancient name continue in today's European languages – *orquidea, orchidea, orchid, orchidée*). His *Enquiry into Plants* describes their roots as having 'a double bulb, one large and one small'. For medicinal purposes, 'the larger, given in the milk of a mountain goat, produces more vigour in sexual intercourse; the smaller inhibits and forestalls'. Dioscorides, a first-century Greek herbalist who may have served in the Roman army as a physician, copied Theophrastus's ideas in his five volumes on the medical properties of plants, *De Materia Medica*, which circulated continuously down the centuries, advising herbalists in Latin, Greek and Arabic. In time, even Christianity incorporated Doctrine of Signatures beliefs into its view of the world (as the botanist Robert Turner asserted: 'God hath imprinted upon the Plants, Herbs and Flowers, as it were in Hieroglyphicks, the very signature of their vertues').[11] So, writing within this tradition, Gerard confidently explained that the 'full and sappy rootes of Ladie traces eaten or boiled in milke, and drunke, provoke venery', and that Dogs Stones 'being drunk . . . stirreth up fleshly lust'.[12]

Plainly, a few centuries ago, Britain's orchids were culturally significant medicines, but at some point they faded out of the cultural consciousness. As recently as the seventeenth century, native orchid roots were used to make a hot drink called 'salep' (also known as 'salop' or 'saloop'). Prior to the widespread consumption of tea and coffee, this was available in salons and street corners across Europe, but, like herbalists and apothecaries, it has largely disappeared.[13]

In the process of finding out more about Britain's native orchids I had uncovered an entire history and, within that history, a mystery: When and why had Britain's orchids faded from our culture? Further sleuthing led me to a selection of influential botanical texts spanning the intervening centuries. I thought they might hold the answer. Thomas Johnson's *Descriptio Itineris Plantarum* (1632) noted that the species known today as Early Purple, Green-winged, Heath Spotted and (probably) Southern Marsh were 'common' in and around Hampstead Heath (at that time a heath beyond the northern boundary of London).[14] Listing dozens of orchid species, John Parkinson's *Theatrum Botanicum* (1640) noted that orchids 'grow in the fieldes of our owne country in divers places', some in particular 'in dry grounds, heaths, and waste untilled places, and the like'.[15] Fifty years later, John Ray's *Historia Plantarum* (1686) observed that a couple of species were rare, but he provided locations where all could be found 'pretty plentifully'.[16] Another fifty years and John Blackstone's *Fasciculus Plantarum Circa Harefield* (1737), which listed the names and locations of plants found near Harefield, a village north-west of London, included fourteen orchid species, observing that they grew 'plentifully'.[17]

The reason orchids appeared in these books was constant too: they were used for medical purposes, primarily to incite lust. Then something changed. In William Curtis's *Flora Londinensis*, published in six volumes between 1777 and 1798, the Bee is not a Foxe stone, nor is it *Orchis Melittias*. Curtis refers to it as *Ophrys apifera*, the name still current today. He gives detailed Latin and English descriptions of each part of the Bee, but the roots get two lines and the flowers thirty-one. Of the Bee Orchid tuber, Curtis notes (referencing 'Salop', the drink more commonly made from the Early Purple Orchid), 'The root appears to possess the same virtues with those of the Orchis from which Salop is made, but being much smaller, is not worth cultivating on that account.' There is no mention of lust.[18]

For Curtis – the director of the Society of Apothecaries – and, it can be assumed, for culture at that time, orchid *flowers* were now more significant than their roots. As the publication's full title makes clear, plants were no longer solely considered in terms of medicine: they had 'several uses in medicine, agriculture, rural economy, and other arts'. Curtis's comments on the Bee reflect this shift: 'The great resemblance which the flower bears to a Bee, makes it much sought after by Florists, whose curiosity indeed, often prompts them to exceed the bounds of moderation rooting up all they find, without leaving a single specimen to cheer the heart of the Student in his botanic excursions.'[19]

In contrast to the findings of earlier writers, by Curtis's time the Bee 'is become so rare about *London*, as scarcely to be found with any certainty'.[20] Florists alone may not have caused this loss. By this time gardeners had started playing a role. Curtis notes, for example, that Bees can be taken from the wild and used as garden plants.[21]

Herbalists continued to ply their trade throughout the nineteenth century, but gradually new medical beliefs gained acceptance and replaced the far older herbalist beliefs about the body, ailments and cures.

It occurred to me that because of the widespread scepticism around modern medicine, and the use of herbal remedies which continued into the twentieth century, changes in medical practice alone were unlikely to have been the most significant driver for the fading of native orchids from culture. As more of the population lived in cities, could it rather be that orchids growing in meadows, marshes, heaths and woods, visited by fewer and fewer people, were simply forgotten: out of sight, out of mind? Perhaps. It was also possible that the gradual erasure of orchids' 'vertues' may have owed something to an increasing sense of Victorian propriety in which explicit mention of venery, lust, ballocks and testicles was deemed inappropriate (at least in relation to flowers). Aesthetically pleasing flowers, blooming in the world of air and light, were far more respectable subjects (even though flowers are, of course, the sexual organs of plants).[22]

The development of modern science might offer another explanation. Victorian scientists (and, increasingly, wider culture) treated nature as separate from the human realm. A commonly held perception was that leaving nature to its own devices would allow it to go wild, untamed, rampant; nature's amorality, godlessness and lack of perceptible progress and order were anathema to the Victorian's self-proclaimed drive towards progress, civilisation, order and Christian morality.[23]

In some ways, of course, this Victorian attitude to the natural world was a remodelling of the older Christian idea that humans were God's special creatures given dominance over the Earth. This Victorian re-imagining vindicated the view that 'civilised' human races could control nature and exploit it, along with any 'lesser' peoples and species. Now science *and* God positioned 'civilised', 'moral' man at the top of a pyramid constructed from the progressively lower, less evolved, less intelligent, less important creatures. As European technological, economic and military power spread, so did these ideas.[24] Nature didn't – and still doesn't – matter like it used to. Could this be partly why Britain's orchids disappeared from culture? Could it also have been that the Victorian mania for exotic orchids eclipsed the older domestic history of the nation's own species with a thrilling, timely, modern story? These narratives involved daring tales of European orchid hunters overcoming nature to rescue orchids from those ungodly foreign places to bring their living treasures back to civilised Europe.[25] By the time Orchid Fever reached its height, a slew of horticultural and botanical studies appeared, many of the most sumptuously illustrated devoted to orchids, but Europe's native species are almost entirely absent from them.[26] This absence may not be solely due to a change in fashion.

Observations made in 1797 by James Smith in *English Botany* reveal the state of native orchids at that time. 'The Bee Orchis', he notes, '[is] generally a favourite with all admirers of plants, and has by that means become rare in the neighbourhood of London and other great towns, having been rooted out by the rapacity of cultivators.'[27] Of the Lizard Orchid, Smith remarks, 'The greediness of collectors has frequently endangered its total destruction, and in some seasons none can be found in flower.'[28] By 1862, Darwin referred to the Lizard as 'extremely rare'.[29]

It seems that decades before wealthy members of British society developed a taste for expensive tropical species, Britain's own orchids had become victims of their beauty. There could have been lessons to learn from their rapid decline; instead the increasing availability of tropical orchids soon obscured their history. Jungles were plundered; hundreds of thousands of tropical orchids were needlessly destroyed.

Native orchids still appeared here and there, but much less in books about orchids than in general accounts of British wild flowers. One of these, *Flora of Middlesex: A Topographical and Historical Account* (1869), lists many species of native orchid, but they are no longer described as growing 'profusely'. The Green-winged has become 'rather rare', Butterfly Orchids, Early Purple and Heath Spotted are 'rare', and the Bee, Pyramidal, Lady, Burnt, Early Marsh, Bird's-nest, Heath Fragrant, and Military are all 'very rare'.[30] Less than a century earlier (in 1777) Curtis referred to the Military Orchid as growing 'in many places in Kent, especially about Rochester . . . in great abundance';[31] in 1929 the last known wild British Military Orchid was thought to have been picked.[32] Species after species, the rapid disappearance of Britain's orchids was obvious, but it was a loss which, with my attention on the felling of jungles and the disappearance of tropical species, I had woefully neglected – as, perhaps, had others who were mesmerised by gaudy tropical species.

The Bees in my father's garden were the remnants of a once-extensive population. Uncovering the forgotten threads of history and culture entwined with them felt like peeking into the world of our ancestors, before internal-combustion engines, plastic and the Internet, when meadows, woods and marshes were pharmacies, and herbalists mixed cures based on ancient recipes. Disappointed that I had been carried along by that same cultural erasure, I resolved that as soon as my back allowed, I would find more living remnants of Britain's orchid-filled past.

4

LOST LADY'S-TRESSES

Sand dunes sweep up to a Mohican strip of marram grass. Below the seaward face of the dunes, the English Channel sifts trillions of sand grains up the beach and down. Half-naked holidaymakers shout and splash and dig and bask, soaking up the seaside atmosphere just like I did years ago, when my parents brought me here with a bucket and spade. Today, while the beach is physically the same, I see it differently. What matters now is the sandy valley behind the dunes.

I first came across the term 'dune slacks' in the Harraps' *Orchids of Britain and Ireland*. It's the name for the dips between sand dunes which are good places to find native orchids. Salt water, sand, North Atlantic winds and . . . orchids? A mind conditioned to tropical species regarded these conditions as the exact opposite of suitable orchid habitat. However, I was starting to learn that Britain's orchids had a great deal to teach me.

My spine wasn't yet sufficiently healed for me to roam the country searching for the magical-sounding Red Helleborine, exquisite Lady's Slipper, quirky Late Spider and cheeky Monkey. For the time being, the distance I could travel was limited. Yet of Britain's fifty-odd orchid species, under a dozen, none of them especially rare, could be found within fifty miles of where I lived. For a new British orchid enthusiast that was plenty to be getting on with. On the other hand, I now knew that my encounter with the Bee had occurred halfway through the annual orchid flowering season, which (in that corner of England) runs from late April to July, followed by the late-autumn flowering of Autumn Lady's-tresses. By the time my spine was fully healed, almost every species in the country would have finished flowering for that year. I would need to wait until the following year to witness the full array of the nation's orchids. Fortunately, there were still a few opportunities to find some accessible flowering species. So it was that as soon as I physically could, I took a twenty-minute train ride to a small seaside nature reserve where, according to the Harraps, I might find Marsh Helleborines.

Until recently I had thought every orchid was called, well, an 'orchid' (Bee Orchid, Military Orchid, Monkey Orchid), but I now knew that Britain's orchids included 'twayblades' and 'helleborines'. I was especially excited about the prospect of encountering a Marsh Helleborine because they were described as 'stunning . . . miniature versions of the gaudy hothouse hybrids'.[1] So Britain had its own tropical-looking orchid? I was about to find out. Being in the big wide world felt good. I was on a mission. And it involved orchids. That was enough to mute the pain of my partially healed back.

From the station I crossed a sand-strewn car park towards a wooden gate leading to the reserve. Wedged between a golf course and the beach, the dune slack was a broad, flat valley carpeted with low-lying wild thyme and petalwort. Maybe a kilometre long, it formed the central slice of a broad sand spit which projected into the estuary towards a town on the opposite shore. I was the only person there. No power lines, no roads, no sound of cars. That marginal land between beach, dunes and fairway seemed to close around me, a mellow piece of England bypassed by time, the perfect location for forgotten plants.

The July sun beat down. Twenty-foot-high dunes deflected the sea breeze. The still air thickened with insect song and the mingled perfumes of hot leaves, sand, nectar. A few reedy pools glinted. Spears of irises and reeds bordered them. Around their thickets of stems I noticed a small forest of thinner stems, each about 30 centimetres tall, hung with lace-edged flowers that were almost white but tinted here and there with a dusky pink. I was pretty sure those frilly colours were what I was looking for. I hurried closer.

My trainers sank into wet sand. I didn't care. I focused on the flowers. About two dozen flower spikes rose around the pool, each wreathed with white-pink flowers. Most of the flowers drooped downwards, but when I lowered myself to their level and peered into one, I saw a five-pointed star, its interior rich with fine purple stripes, blobs of gold and green, and a broad white lacy lip, each white petal washed with green, pink and mauve. No doubt about it, they were Marsh Helleborines, or *Epipactis palustris* – *Epipactis* probably dating all the way back to the name Theophrastus gave to a plant used to curdle milk and then applied (in the eighteenth century) to these orchids because of their similar appearance; *palustris* meaning 'of swampy ground'.[2]

Probably first recorded in English in 1633 in an expanded edition of Gerard's *Herball*, as with other helleborines, several differences set them apart from other orchids. Flower structure is one of these. In helleborines the large flower ovary seems to form the flower's stem, and the flower's lower lip consists of two parts. This is quite different from the flowers of the Bee.

A little awestruck, I breathed the helleborines' heady vanilla scent, then rested in the insect-humming heat. A plan was forming. Judging by that trip's success, finding native orchids was simple. The forgotten places of England had to be rammed with them. All I needed to do was track down some other similarly preserved scraps of England and I would find orchids. I resolved to make that my mission: I would photograph every different species of native orchid in the land. Even with my limited ability to travel, I had already found two. Surely it would not take long to find more.

★

Much of the south Devon coastline looks as if an immense Jurassic sea monster gnawed on it long ago, its huge jaws leaving hundreds of chomp-shaped, cliff-backed coves. One bite forms a broad crescent between Hope's Nose in the north and Berry Head in the south. This is Torbay, the destination of my next orchid-hunting expedition: an area of field- and wood-backed beaches, forts built to protect the area from Napoleonic invasions and seaside towns tumbling down to beach huts, fish-and-chip shops and boats bobbing in harbours. Canary Island palm trees and New Zealand cabbage trees grow there, lending an exotic flourish; the climate is mild and on bright days the sea is a holiday-brochure azure, but, for the same reason, high season sees shrill crowds pack the beaches and gridlock grip the narrow roads.

I was heading to a cliff-top section of this English Riviera, a reputed hotspot for orchids, where Pyramidal, Early Purple, Green-winged, Bee, Common Spotted, Common Twayblade and Autumn Lady's-tresses could be found at the right time of year.[3] The route to the headland was a potholed single-track lane which dipped and curved through woodland. I followed it between high banks shaggy with ferns, mosses and liverworts, while beech, chestnut and whitebeam formed a dappled green roof. No street lamps. No grumble of engines. No one. Only birdsong. Insect hum. Unseen waves whispering. If it hadn't been for the broken tarmac, I could have slipped once again to almost any moment in the past ten thousand years of English history. As the tree tunnel closed overhead, I reached for my camera, ready to snap the orchid I most wanted to find: the Common Twayblade.

Gerard described the Twayblade's flowers as 'resembling a gnat, or little gosling newly hatched' (exactly how similar gnats and newly hatched goslings are perplexed me) and observed that its medical use was not for encouraging 'venerie', but 'for greene wounds, burstings, and ruptures; whereof', Gerard claimed, 'I have in my unguents and balsams had great experience, and goode successe.'[4]

The Common Twayblade (*Neottia ovata*) flowered between April and August. It was the second half of July. The timing was auspicious and one of the Twayblade's favoured habitats, moist, shady, deciduous woodland, rose all around. It wasn't just the promising alignment of time and place that made me intent on that particular species. The plant itself seemed to be rather strange; that strangeness made it attractive.

The flowers were tiny, delicate five-pointed stars, the lowest two points somewhat elongated, but an inconspicuous light green, the same shade as the stalk, as grass, as a host of other leaves. They were far from the colourful blossoms I associated with orchids. I'd learned it can take up to twenty years for Common Twayblade seed to become a flowering plant, and its seedlings spend three or four years completely underground.[5] How was that possible, I wondered, given that plants need sunlight? What's more, Common Twayblades can live a long time: their 'half-life' (the timespan between the first recording of the number of plants in a colony and half of the original plants dying) has been recorded as over seventy years.[6]

All that – a low-key English perennial, plain, common, an orchid not called an orchid, with an extraordinary ability to live underground and to a ripe old age – made everything about the Common Twayblade either contrary to all that orchids usually stood for (at least as I had come to understand orchids in terms of tropical ones) or completely remarkable. I was determined to find one.

Chestnut saplings jostled with hart's-tongue ferns in the tree-muted shade. Liverworts, wood vetch and purple gromwell (a nationally rare plant in its own right) cloaked the earth. Slowly, I made my way along the margins of the lane. Not one glimpse of the Twayblade's twin oval leaves. No flower spikes covered with green gnats or goslings. I was frustratingly aware that Twayblades could be flowering in the dense wood, obscured by undergrowth, or that hundreds of seedlings could be nestled in the earth, but I had no way of knowing. Or perhaps Common

Twayblades were simply too inconspicuous for a native-orchid-hunting rookie like me.

My heart sank as I reached the end of the lane. I considered retracing my steps and searching again, but a whole headland was waiting. Other species that preferred open grassland, such as Pyramidals, could be there.

I came to a small, empty gravel car park. A faded sign declared the area to be an SSSI – a Site of Special Scientific Interest. 'Fires and Camping Prohibited.' Billowing ranks of buddleia, stinging nettles and bramble pressed in on either side. I chose the flattest-looking path and shuffled onto a rise of short turf.

A salt breeze rippled the broad expanse of grass scattered with daisies. A skylark poured its song from somewhere overhead. To my right, hedgerows marched up gentle slopes apportioning cow-grazed fields. Ahead and to the left crumbling shoulders of land dropped to fingernail crescents of sand lapped by the shiny blue-green-grey Channel. A few low, dark cargo ships floated where the sea merged with the hazy sky.

I turned my attention to the grassy headland and what was growing there. The path threaded along the cliff top. Head bowed, I scanned the vegetation, looking hopefully for the lilac, pink, white and black speckled spires of Common Spotted Orchids or the pink pyramid-shaped towers of Pyramidals. The path meandered through clumps of bracken and bramble. No orchids. I circled the headland, scanning every inch. I leaned as far as I dared over cliff edges, attempting to spot the pink splash of an orchid clinging to the crumbly earth. There were no pink splashes. Golden samphire grew in cheerful yellow clumps. Thistles offered purple crowns. I searched and circled. No orchids. I peered at the other meadow flowers, numb, puzzled. My hopes ebbed away.

Every speculative tropical jungle exploit I had embarked on had offered at least one flowerless orchid. Yet, in my own country, orchids had eluded me in a location renowned for their presence.

What was I missing? Had I arrived a week or two early? Too late? Perhaps if I walked to the brow of the nearest hill I would find carpets of orchids. Hungry, back aching, I considered the grazing cows, the five-bar gates, the hedgerows. With the promise of the hill, the cliffs and the headlands receding into the hazy blue distance, it was time to cut my losses.

In dour mood, I shuffled back past the billowing buddleia and brambles to the tree tunnel of the lane. I slowed, wishfully checked the undergrowth. There was still time for the Common Twayblade to save the day. It did not.

After years of successfully finding tropicals, I wasn't prepared to accept I was *that* inept at finding orchids. It occurred to me that the disappearance of Britain's orchids was more than cultural. They had slipped out of the landscape too. At that stage, the suggested explanations I had come across for their loss from the land were the 'rapacity of cultivators' and the 'greediness of collectors and Florists', but it seemed implausible that (unlike the Bee's remarkable blooms) the supposedly commonplace and modest Common Twayblade had ever been popular with collectors, gardeners or florists. So, if *they* were not responsible for its disappearance, what or who was?

Trundling back home on the train, I took out my *Orchids of Britain and Ireland* and ruefully studied the photos of the plant I had failed to find. I reread the text. 'Common Twayblade', the Harraps conceded, 'has vanished from almost 30% of its historical range in Britain and Ireland, with a relatively large proportion of the British losses being recent.' In my eagerness I had simply mind-blanked that line. The words 'losses being recent' struck me as curious.[7]

My notion of a 'Green Britain' was supported by years of schooling in which I had been taught the ills of slash-and-burn farming, the ozone hole and the greenhouse effect; years of the family television transmitting a diet of virgin rainforest being felled, burnt and turned into charcoal deserts; years of the media, celebrities and politicians condemning the damaging effects of acid rain, the ivory trade, whaling, egg-collecting and the destruction of rainforests. Government after government had pledged to cut greenhouse-gas emissions, make disposable plastic and plastic microbeads illegal, create more protected natural spaces, reduce landfill, increase renewable energy sources and on and on. Furthermore, Britain is home to some of the world's oldest publicly funded nature-conservation organisations. The National Trust, founded in 1895, the RSPB founded in 1889; the Botanical Society of London (now the Botanical Society of Britain and Ireland) in 1836; the RSPCA (founded in 1824); the Royal Horticultural Society (in 1804), and the Royal Botanic Gardens at Kew, formally founded in 1840 but with origins going back to 1772, an internationally significant institution, highly regarded for its research into and conservation of fungi and plants (with an emphasis on orchids).[8]

For well over a century (or two), all those royal endorsements and efforts of charitable public support were testament to Britain's environmentalism and its eco-savvy population. Clearly – so I believed – Britain and Britons knew how to respect the natural world; surely all those environmentally destructive activities – soil erosion, air pollution, run-off of nitrates into waterways, loss of virgin habitat – occurred in *other* countries?

If only the missing orchids on that protected headland told the same story. And it wasn't just Twayblades. I discovered that many native species had lately gone extinct in my part of the country. The Fen Orchid, with flowers like skywards-facing lime-green whimsical stars, now grew in only a handful of tiny areas of Britain: a few fens in Norfolk, one site in South Wales and, until 1987, in North Devon. The Burnt Orchid disappeared from the county in the 1930s. The Narrow-lipped Helleborine, Lizard Orchid, White Helleborine and Fly Orchid existed in Devon at the beginning of the twentieth century, but had disappeared by the 1950s.[9] The only verified location where Irish Lady's-tresses existed in England was in Devon, but the plants had not appeared since 1993. With their loss, Irish Lady's-tresses had died out in the country.

Those Devonshire losses were part of a pattern. Orchid species had gone locally extinct in counties across the nation – the Lesser Twayblade was extinct in Shropshire, Derbyshire, Cheshire; the White Helleborine in Nottinghamshire, Shropshire, Derbyshire, Warwickshire and Essex; the Violet Helleborine in Cambridgeshire and Norfolk; Creeping Lady's-tresses in West Lothian, Dumfriesshire and Orkney. The Lady Orchid disappeared from Surrey in 1959, Herefordshire in 1967, West Sussex in 1976 and Somerset in 1990. Populations of the Small White had collapsed in Wales and it was extinct in Kent, East Sussex, Gloucestershire, Staffordshire and Lancashire. The Lesser Butterfly can no longer be found in Essex, Suffolk, Middlesex, Worcestershire, Derbyshire, Berkshire. This list could continue (see the Appendix for more detail on Britain and Ireland's lost orchids), and that's without mentioning the elusive Ghost Orchid (*Epipogium aphyllum*), last seen in Britain in 2009. No one can (yet) be certain if it has gone for ever. If it has, it has gone the way of Summer Lady's-tresses (*Spiranthes aestivalis*), last seen in the UK in the 1950s.

My interest in Britain's orchids might have come too late to see the ethereal squid-like white-pink, banana-scented flowers of the Ghost emerging from British earth. That seemed wrong. Shockingly wrong. I was suddenly waking up from comfortable ignorance. I thought this nation was environmentally friendly, a nation that (as a slogan adopted by the RSPB says) gives 'nature a home'. That story had guided me to travel the world to see tropical orchids before their forests were torched or felled, but Britain's orchids were revealing a tale of accelerating loss from *this* land.

Governments, media, charities, public opinion – all widely condemned the felling of rainforest in other countries, but in 2008 research showed that more than half of Britain's remaining 'ancient' woodland had been lost in the previous fifty years. That was a rate faster than the loss of the Amazon jungle; only 2.5 per cent of the woodland in Britain is ancient and native, and much of that is in a poor condition.[10] Over about the same period, around 80 per cent of the UK's peatlands have been lost and 97 per cent of England's wild-flower meadows vanished.[11] These lost habitats have had a direct impact on orchids.

'Watery middowes and in woddes', 'moist and fertill meadowes', 'pastures and fields that seldom or never are dunged or manured' are among the habitats favoured by orchids. There, herbalists found them growing 'profusely'. With loss of habitat, comes loss of species.

To stem this loss, in the 1990s, the UK government established a Biodiversity Action Plan (BAP) overseen by an advisory body called the Joint Nature Conservation Committee (JNCC). The first of its kind in the world, the BAP seemed to confirm the UK's ecological values. Between 1995 and 1997, the JNCC compiled a UK Biodiversity Priority Species List of 577 threatened species whose populations it deemed deserved attentive monitoring. In the subsequent twelve years that number of priority species – from the hairy click beetle, cirl buntings, water voles and slow-worms to wild asparagus and leaf-scraper sharks – doubled to 1,150.[12]

In due course the Biodiversity Action Plan changed its name to the Priority Species List and a UK-wide list was divided up into a list for each of the four countries. Essentially these lists serve the same purpose: they are a tool used by the JNCC to provide the UK government and devolved regions with 'evidence and advice on nature conservation and natural capital'.[13]

As one of its key indicative measures the JNCC plots a graph of these species' populations over time. Somewhat randomly (if not inaccurately; by then of course one species of orchid was already extinct in Britain), it uses 1970 as the year to represent the point at which populations of all priority species were at 100 per cent (though this may have represented only 50, 20 or

10 per cent of their populations present in 1700 – no one really knows).[14] Even with this questionable starting point, the JNCC graph shows that despite being a much-vaunted government initiative, between 1970 and 2018 Britain's priority species populations slumped as if from a lofty peak to a murky valley.[15]

The JNCC is keen to point out that by 2007, although many species had been added to the priority species list, 123 had been 'delisted' because they no longer met the criteria for inclusion. They explain that 'in many cases this was due to conservation action'.[16] True, of the 123 species, nineteen were delisted as the result of 'targets' being met and 'recovered' populations. Four others were delisted because they had become or were already 'extinct', with 'no plans or realistic possibilities for reintroduction', and twenty-two others were cut due to 'taxonomic revision' or confusion with a different species. The data source gives no explanation for why dozens of other species were removed.[17]

Learning all this, I felt as if I had been duped. Orchids had exposed Britain's much-vaunted eco-credentials for what they really were. The reality was that, ranked out of 180 countries on the Yale University Biodiversity Habitat Index, the United Kingdom comes in equal 143rd (alongside El Salvador and Bangladesh).[18] It has even been claimed that the United Kingdom has 'led the world' in destroying the natural environment and the wildlife which lives there.[19]

Today's Priority Species List includes fourteen orchids: the Lady's Slipper, Irish Lady's-tresses, Fen, Man, Frog, Musk, Fly, Monkey, Burnt, Lesser Butterfly, Small White, and the Red, Narrow-lipped and White Helleborines, as well as two subspecies of Marsh Orchid. My chances of personally encountering any of those orchids were diminishing by the day. The same feeling of a race against extinction that had part-fuelled my search for exotic tropicals returned, this time directed at those species I had overlooked for so long.

Tropical orchids were disappearing because of habitat loss, but I now knew that British species were too. Sometimes this was clearly visible habitat demolition, essentially the slash-and-burn we disapprove of in other countries, just without the fire. This loss owed much to urban sprawl and a post-Second World War government drive to increase Britain's agricultural output.[20] The result? Little-used areas of the countryside were claimed for intensive farming; marshes, fens and bogs were drained and uncultivated land was ploughed and planted with swathes of single-species crops. Financial incentives for farmers made efficiency and output paramount. As mechanised farming increased, less productive traditional farming practices, such as leaving land fallow or setting it aside for hay meadows, were discarded. Herd numbers grew. Soil nutrients, unable to recover naturally, had to be bolstered with artificial fertilisers. As a result, a second wave of less visible habitat destruction occurred.

Crop production benefited from artificial fertilisers and generous helpings of synthetic biocides (herbicides, fungicides and pesticides) to control the pests that could easily run amok through one big area of a single species of plant.[21] The problem was – is – that these additives are not precision tools. Insect pests die, but so do useful pollinators and the natural predators of insect pests. Wind blows sprayed poisons (causing 'drift') away from its intended agricultural targets and onto surrounding areas. As a result, wild flowers and important soil-dwelling fungi die. (For reasons I will come to later, there is often a direct correlation between the health and concentrations of certain soil fungi and the health and prevalence of terrestrial orchids.)[22] It's not only wind which spreads these man-made poisons: animal waste often contains worming products, antibiotics and other artificial chemicals.[23] The waste breaks down on the field surface releasing chemicals into the earth. Rainfall carries artificial fertilisers and biocides through soil and along water courses, changing the chemical composition of the earth as it goes, resulting in plant extinctions, loss of beneficial fungi and increased concentrations of toxic nitrogen hydroxide (with its detrimental effects on aquatic life).[24] Orchids, like so many of Britain's wild flowers, had evolved to inhabit naturally infertile soil full of certain crucial microbes – a habitat destroyed by modern commercial farming.

Nitrification, loss of pollinators, the effects of herbicides: I was starting to understand the degree to which Britain's orchids were succumbing to visible *and* invisible habitat loss. That wasn't all. When was the last time most people in Britain today milked a cow? Tilled a field? Coppiced a wood? Scythed a meadow? Sheared a sheep, spun and wove the wool? Trapped and skinned a rabbit? Set a net in a river for fish? Ventured into the fens to harvest sedge to thatch their house? Going back only a few generations and from there thousands of years, that was life for our ancestors. By using ox-drawn ploughs, spades and axes, grazing animals, planting hedges, digging ditches, forming dykes, managing woodland, mining and hunting, they shaped the fabric of the land.[25] This landscape, formed and maintained by people, benefited native orchids. Heath Spotted, Heath Fragrant, Green-winged, Lesser and Greater Butterfly thrived in meadows where the vegetation was cut once or twice a year and grazed by small herds of livestock in the autumn and winter to keep less useful scrub at bay. Military, Lady, Monkey and many helleborines boomed in woodland and woodland clearings, where coppicing and pollarding thinned the canopies, allowing light to reach orchids growing at ground level for some seasons. In other years, by gradually increasing shade or by the woodland grazing of livestock, grasses and other faster-growing species that would out-compete orchids for light were kept in check.[26] Man, Frog and Fly prospered in the alkaline conditions of disused limestone quarries. On short turf and heathland, where small herds of livestock grazed long grass, brambles, saplings, bracken and other competing species, Lesser and Greater Butterfly, Green-winged, Frog, Musk, Small White and Burnt found the conditions they needed: light, and well-drained, infertile earth.[27]

After millennia of a closely intertwined relationship between people and the natural world, industrialisation heralded huge change. People moved away from agricultural work and rural locations. Management of woodland declined. Plastic goods and prefabricated furniture replaced the need for articles handcrafted from local underwood harvested from deciduous woodland. Many of those woods were grubbed out and replaced with fast-growing non-native conifers to satisfy demand for construction timber.[28] Plantations of evergreens don't offer the fluctuations of light and dark that woodland orchids need to survive and they lack the beech and oak with which some orchids have a crucial relationship (more on this later). Moreover, heavy forestry machines churn up the earth and with it these slow-growing plants.

Of course it is not only woodland that has suffered. Fens, marshes and bogs were drained to make way for housing, or water was extracted to irrigate huge fields of crops. No one had time to scythe hay meadows. No one bothered to rake up the cuttings from mown grass verges. Once upon a time grass cuttings were useful food for livestock. Today, cuttings decay where they fall, increasing the soil's nitrogen content. Deer and rabbits used to be hunted and, by this means, their numbers controlled. In areas where their populations have exploded – including in nature reserves – plants are grazed before they can set seed. Conversely, where rabbit populations crashed (after the myxoma virus was illegally introduced to Britain in 1953) grasses and scrub grew profusely, smothering less robust species.[29] On top of all that, society's mounting quantities of rubbish needed to be piled somewhere: abandoned quarries and chalk pits offered an easy solution.[30]

Every factor listed above has contributed to the loss of Britain's orchids, but clearly, this is not always a simple tale of natural innocence obliterated by unthinking people. Consider that Site of Special Scientific Interest: at first glance, pristine, protected from urban encroachment, left to nature and orchids, yet . . . orchidless. Like much of Britain's countryside, those cliff tops had not really been 'natural' for thousands of years. The decision to designate it as an SSSI and to 'conserve' and 'protect' it – by preventing people and animals from clearing and using it – had neither conserved nor protected it. The short turf had been exposed to encroaching scrub – bracken, buddleia, thick bramble tendrils. For centuries kept in check, allowed to grow free they had taken the opportunity to invade, in the process starving orchids of light and pollinators. In the previous decade, as a result of that well-meaning but misguided wish to conserve a little bit of 'natural' England by putting fences around it, the last orchids there had died out.

So it was that orchids flipped my ideas of conservation.

I had resigned myself to waiting until the following spring to see more local orchids when, one day in October, I happened upon a home-made sign – 'Caution! Wild Orchids!' – staked into the short grass of a roadside verge. I took a closer look. The sign had a photo showing a flower spike, green and twisted as if plaited from tendrils of living grass and hung with a helix of pale, delicate little trumpet-like blooms.

In front of me, emerging from the short grass, their slender flower spikes quivering in the slipstream of passing vehicles, were about twenty Autumn Lady's-tresses. I wondered who had erected that sign, but this question was quickly superseded by the pleasure of accidentally finding my third species of native orchid, not in a distant reserve, private garden or slip of land that had dodged modern times, but beside a busy road used daily by thousands of people.

I crouched beside them, attracting odd looks from passing drivers. I didn't care. There was something irreverent and resolute about those tiny orchids buffeted by the diesel breath of passing trucks. I had read so much about the disappearance of orchids, but these diminutive beauties proved nature *could* coexist with the modern world.

I returned the following day, camera in hand. The sign had

fallen over; the plants were unharmed. I stuck the sign upright and spent a while trying to get the best angle on those impertinent flowers. Crouching there, I felt the way I had when photographing tropical species abroad, which is to say, I had a sense of cataloguing something threatened with destruction. Indeed, that was what I was doing.

In 1964 the International Union for Conservation of Nature (IUCN) categorised the conservation status of all the planet's wildlife. The Red List of Threatened Species remains a keystone of conservation policy. The categories extend from 'Extinct' and 'Extinct in the Wild', through 'Critically Endangered', 'Endangered' and 'Vulnerable' (all considered at risk of extinction), to 'Near Threatened' and 'Least Concern'. Autumn Lady's-tresses are classed as Near Threatened. In Britain, they have vanished from 55 per cent of their former range.[31]

As I left that roadside orchid colony, I noticed an advertising hoarding announcing the development of the neighbouring fields for housing. I thought little of it. There were many such signs.

Fast-forward to a dreary midwinter afternoon, sunlight struggling to heal a badly bruised sky. I drove past the same verge. I looked out of the rain-beaded window to check on the turf where those orchids grew. There was no grass. There was mud scarred by huge tyres. I pulled over. That mire was indeed where the orchids had been. The hedge had been uprooted and replaced with steel fencing. The fields beyond had become a red-brown wasteland of naked earth hacked into drainage ditches and foundations. Through silver scratches of rain I saw stacks of massive concrete pipes, orange flags marking out some kind of boundary, buttercup-yellow construction traffic. An excavator perched on top of a pyramid of earth like a victorious, one-clawed dragon. Where were the orchids?

I had wanted to catalogue loss; I had not imagined that loss might come so soon. In that moment, almost before it had begun, I realised the futility of my latest orchid-related project. I had photos of the now-gone orchids, but what good were they? The JNCC, the IUCN and scientists worldwide were all carefully tracking the loss of species, but they weren't stopping it from happening.

A call to the local planning authority offered little reassurance. I was politely informed that the period for public consultation for that development had ended some time ago, but was assured that 'net biodiversity gain' was a key element of the strategic plan for the whole metropolitan area. The planning authority had carefully considered the developer's environmental impact statement alongside measures to address any environmental damage caused during the construction process. I asked if that involved reinstating orchids. The planning officer couldn't say without checking the relevant documentation, but he suggested I could do that myself: all the paperwork was available via the county council website. I thanked him and opened my laptop.

EXTINCTION'S CORRIDORS

The county council planning portal introduced me to a labyrinthine world of maps, plans, photographs, core development strategies, geotechnical assemblies, minutes of meetings, letters from the public, studies of cultural heritage, unemployment rates, water and earth quality, transport links, pollution forecasts and much, much more.

I ploughed through the documents for the development but found no mention of orchids. Bats had been considered, as well as badgers, otters, great crested newts, toads, slow-worms, dormice and hedgerows. They were surveyed, they had reports written about them, and all those reports shared similar concerns: the development would destroy foraging, dispersal, breeding, commuting and hibernacula habitat for local species. The subsequent increase in human activity would degrade the diversity of ground flora; the introduction of domestic cats would increase predation of protected species; artificial light had a good chance of altering the diurnal cycles of local species; and two ponds on the site, although preserved within the development, would be put under increased pressure due to human activity, pollution and littering.

Despite these observations, and without mentioning orchids, the development had got full council approval.

Wondering whether I had uncovered my very own *Pelican*

Brief-type cover-up, I decided to take a closer look at the law, in particular the law around wild plants.

For decades legal protection for every wild mammal, insect, bird, plant, lichen, fungus, moss and fish in Britain has derived from the Wildlife and Countryside Act 1981. Section 13, 'Protection of wild plants', was most relevant to my investigation. Section 13 opens by stating it is an offence if an individual 'not being an authorised person, intentionally uproots any wild plant'. So far, so good. This seemed designed to discourage people like those eighteenth-century gardeners and collectors who caused so much damage to Britain's orchids by digging up wild plants.

I read on. Uprooting any wild plant is illegal, but picking a wild flower is not. This seemed a bit self-defeating. Flowers produce the seeds needed to start new generations of plants; picking them destroys the source of seeds. I wondered whether the law around 'uprooting' was so specific that *crushing* wild flowers with two-tonne machines is not illegal. As it turned out, by the letter of the law, crushing plants with an excavator or pulverising them with a golf club (or any other object) is not. However, I was relieved to find this is not true for all plants.

The Wildlife and Countryside Act offers greater protection to a select group of species named after the section of the Act in which they appear: 'Schedule 8'. Schedule 8 includes dozens of Britain's rarest plants: among them, eleven species of orchid. The law is very clear about plants on this list: it is an offence not just intentionally or 'recklessly' to uproot, but also to pick, destroy or offer for sale the plant or any part of that plant, alive or dead; to possess, transport and/or advertise for sale any plant or part of a plant (alive or dead), including the seed. This all sounded pretty robust.

I took a closer look at the Schedule 8 orchids. They included the Critically Endangered Lady's Slipper (*Cypripedium calceolus*) and Red Helleborine (*Cephalanthera rubra*), as well as the Endangered Fen Orchid (*Liparis loeselii*) and Vulnerable Military

(*Orchis militaris*) and Monkey (*Orchis simia*), each of these species found at no more than a handful of sites in the whole country.

Oddly, many of the nation's most recently depleted orchids, including the Burnt Orchid (which has suffered a 75 per cent loss of its previous range in under a century), the Musk Orchid (lost from 69 per cent of its former range), the Small White (over 65 per cent), Lesser Butterfly (64 per cent) and Bog Orchid (61 per cent), not to mention *Dactylorhiza ochroleuca*, the Critically Endangered Early Marsh Orchid variant, were completely absent. Autumn Lady's-tresses, despite Near Threatened status (and a recent loss of over 50 per cent of its historic range), also did not make the cut.

What's more, two so-called orchid species which hadn't been considered species for some time – Young's Helleborine and the Lapland Marsh Orchid – made the Schedule 8 list.[1] So the highest level of legal protection was afforded to two orchid species that, technically, never existed (or, in the case of the Lapland Marsh Orchid, is the name of an orchid native to Scandinavia, not Britain), but not to a number of species, including one Critically Endangered variety, that were arguably worthy of inclusion.

These significant shortcomings aside, the law itself seemed quite strict. It even went after anyone who 'knowingly causes or permits to be done' any of the aforementioned intentional or reckless acts involving Schedule 8 plants. Surely that meant that construction companies, site managers, even planning officers who approved the work, were culpable.

I studied the Act again. It largely excuses anyone from causing otherwise unlawful damage to any plant, as long as that unlawful damage is 'the incidental result of a lawful operation or other activity'. Doubtless farming, forestry work and an officially sanctioned construction project fall well within the bounds of a 'lawful operation'. As for 'other activity', well . . . that seemed to leave the door flapping in the breeze. Music festivals? Mud-wrestling? Alpaca wrangling?

On top of that, the Act added that if the individual 'did not foresee, and could not reasonably have foreseen, that the unlawful act would be an incidental result of the carrying out of the lawful operation or other activity', they would not have committed an offence. So, if a survey carried out by the developers did not highlight the presence of orchids on a site, then any damage to that colony would be classed as 'unforeseen' and therefore not unlawful.

Given that for several months of every year and entire years of their young lives, orchids are invisible underground, the chance of 'unforeseen' damage to them is high. In fact, the environmental impact statement for that development (it seems to be standard wording in any planning application) included a paragraph catering directly to that section of the law: 'No account can be made of the absence of a species during the field surveys, and therefore it is possible, although unlikely, that a protected or otherwise notable species may have been overlooked or not recorded on any particular day.'

The vegetation survey for the site had been conducted on one day in May. Autumn Lady's-tresses don't flower until August at the earliest and, until their flower spikes spiral out of the earth, they are easily overlooked and readily confused with other, commoner plants.

More legal caveats followed. If the individual 'who carried out the unlawful act took, immediately upon the consequence of that act becoming apparent, such steps as were reasonably practicable in the circumstances to minimise the damage to the wild plant in relation to which the unlawful act was carried out', they definitely would not be guilty of a crime. Basically, if a lawfully employed excavator driver clears land and mows down orchids and then an orchid geek like me complains, I can grumble all I like but there is no legal case.

On the face of it, the law protected England's wildlife and countryside, but the only people who could really be prosecuted for uprooting or damaging orchids were unauthorised individuals intentionally digging them up (or taking parts of them) for possession or sale. It's true that florists, gardeners and botanists collecting specimens had caused significant wild orchid loss over a century earlier, but were they responsible for the recent loss of 75 per cent of the country's Burnt Orchids? There was little evidence of that. Urban sprawl, cultural change, modern conventional agriculture – *they* are the main culprits. Yet, according to the Wildlife and Countryside Act, these are 'lawful operations'. In essence, the loss of habitat and orchids was sanctioned by law. Knowing this, it was unsurprising those pint-sized Autumn Lady's-tresses had met the fate they did. The law was a deterrent in name alone.

I found much the same contradictions in the National Planning Policy Framework – the document which lays out expectations for the country's construction projects. It foregrounds the importance of sustainability, which it defines as 'meeting the needs of the present without compromising the ability of future generations to meet their own needs', then makes concessions such as allowing development rights even on the rarest 'priority' natural habitat and accepting 'loss or deterioration of irreplaceable habitats' in cases 'where the public benefit would clearly outweigh the loss'.[2] Combined with seemingly ineffective legislation, it all struck me as a kind of game in which professional players can use the rules to their advantage.

Although the orchid colony had been ignored, how the development treated the slow-worms which lived there and the twenty-five hedgerows which crossed the site seemed illustrative of that kind of game played by developers, laws and planning policies with the natural world.

The developers commissioned a report on reptiles and amphibians on the site which stated that digging up those fields and concreting them over would be 'moderately adverse' to slow-worm habitat. The developer's response was that, because urban gardens provide suitable habitat for slow-worms (a JNCC priority species), overall the amount of suitable habitat for this species would increase, meaning that the development would be 'long-term minor beneficial'. Compared to acres of pastureland, I was left wondering how thousands of homes, driveways and roads could reasonably be considered as *increasing* the amount of suitable slow-worm habitat. What was to stop the new owners from paving over their gardens? I was also left wondering how – given there was no mention of rescuing resident slow-worms from the fields and reintroducing them to the site – the developer expected slow-worms to survive heavy construction vehicles razing their entire habitat before merrily wriggling their way into those paradisiacal garden habitats without being crushed by residents' cars, eaten by cats and/or dogs or accidentally being sliced in half by lawnmowers?

As for the hedgerows, the law required the developers to commission a hedgerow survey; it found that single hedgerows on the site contained up to eight tree species. An approximate rule of thumb known as Hooper's Rule states that the age of an English hedgerow can be estimated by counting the number of woody tree species growing along a thirty-yard stretch and multiplying that number by 110.[3] This suggests that some of the hedges could have been part of the landscape since well before the first mention of an orchid in English. Neither the survey nor the developer's environmental impact statement noted this.

The hedgerow survey for that site recognised its hedges as 'dense and species rich'; their removal 'would have an adverse effect on the conservation status of this priority habitat which is certain to be significant at a city level'. Fourteen species of woody plants grew in the hedges, mainly elm, hazel, hawthorn, elder and ash, with lesser numbers of willow, spindle, oak, blackthorn, holly, cherry, field maple, alder and rose. The developer flagged that, at some point, Dutch elm disease was likely to destroy the elm, so it was suggested that 350 square metres of hedgerow might as well be swiftly removed. No comment was made regarding the other species, in particular the large amounts of hazel ('the most seriously threatened British tree except elms'),[4] or the oaks (a mature English oak being home to nearly 300 species of insect and a cornerstone of birdlife).[5] The possibility that some other tree species might grow into spaces left by dead elms was ignored. Only the presence of dormice in certain hedges (at the time of the dormouse survey, over a year before construction began) seemed to matter. However, as dormice weren't present, it was not an 'important' hedge and hedgerow-removal licences were duly granted.

Oliver Rackham described the landscape as 'a historic library of 50,000 books'.[6] From my perspective, the developers and the local planning department didn't share that view. If those hedges had been listed buildings, scheduled monuments or archaeological remains, the law would have intervened, but surely those hedgerows *were* cultural assets, ancient iconic elements of England's quintessential patchwork fields. They were living heritage, maintained for generations – perhaps as far back as when our forefathers believed witches rode up and down hedgerows, the green ways of hedges being visible both in the supernatural witch realm and the physical human one, proof that the hedges offered protection across both planes.[7]

In place of that 'historic library', the developers promised to plant a hedge using native species elsewhere on the development. It didn't seem a fair exchange. It didn't involve much conservation or preservation, and I wasn't sure how effectively it benefited biodiversity.[8] After all, where were all the wild residents that lived on the site supposed to go while their habitat was destroyed?

It appeared, however, that those suggested forms of mitigation were acceptable to the local planning authority and the government's planning regulations. With the tacit support of environmental and planning laws, the minor benefits of a promise to plant a new hedge seemed enough to compensate for what experts had categorised as 'long-term major adverse' effects on habitat, species and environment. So it was that 2,300 houses, small gardens, roads and driveways were approved.

If a reptile on the Priority Species List and hundreds of metres of clearly visible – probably ancient – hedgerows could be removed, what chance was there for diminutive Autumn Lady's-tresses? There was evidence neither that anyone had spared them a glance nor that they had been relocated. Nevertheless, their ghosts had introduced me to a process that seemed to casually allow ongoing extinctions. It was a system that knew how to satisfy *and* exploit laws designed to protect habitat and biodiversity. A planning authority may ask developers to come back with improved measures for protecting hedges, trees or ponds, and they may be advised to take further miti- gating actions such as installing bat roosting boxes. Yet no matter how ecologically minded individuals in planning or construction might be, if the laws that guide them have a reductive defin- ition of 'heritage' and an incorrect list of protected orchids: if they allow field surveys to miss protected species, excuse construction workers for destroying rare flora, regard the replace- ment of ancient with newly planted habitat as an appropriate solution and focus more on economic than environmental benefits – then those who are part of that system are going to play by the rules of the system.

My venture into environmental legislation and its intersection with big-development planning had left me deflated, but amid the twisted logic and game-playing, I came across something interesting. The government's definition of sustainability, about 'meeting the needs of the present without compromising the ability of future generations to meet their own needs', came from a United Nations resolution made in 1987. As useful and accurate as that definition was, when I looked at that UN document, a different line caught my eye. It read, 'The safe- guarding of species is a moral obligation of humankind.'[9] I found myself wondering, in the decades since that was written, who in all of humankind really acted on that 'moral obligation'? Over those decades, all across the planet, Britain being one of the worst-affected countries, biodiversity loss had accelerated.

Of course people need jobs, food and houses, but for millennia orchids – along with much of the 'natural' world including meadows, marshes, woods and hedges – were part of our culture and heritage. Those Autumn Lady's-tresses on the verge by a busy road had been proof that orchids and, by extension, all manner of flora and fauna *could* exist alongside people in a space created and used by people. Part of the problem was that the law failed them.

What happens when the law fails to protect the innocent? What happens when innocent victims have no voice and no justice? What happens when policy-makers ignore them?

The Harraps' book mentioned becoming a 'local champion' for orchids by 'getting to know an area, finding and recording orchids, and then badgering local councils, wildlife trusts, government agencies or church-wardens to sit up and do what is necessary'.[10] I liked the idea, but surely 'badgering' would not be very effective if the law didn't enforce what was being asked. And what if churchwardens, government agencies and local councils didn't want to 'sit up and do what was necessary'?

Could I quietly take 'local champion' to another level? How difficult could it be to monitor planning applications, make trips to sites before they were built on, dig out any orchids I found and relocate them? I now had access to the planning portal to do that. But . . . I would be an 'unauthorised' person intentionally uprooting wild flowers. I would find myself on the wrong side of the law . . . for a moral cause, isn't that sometimes what is necessary?

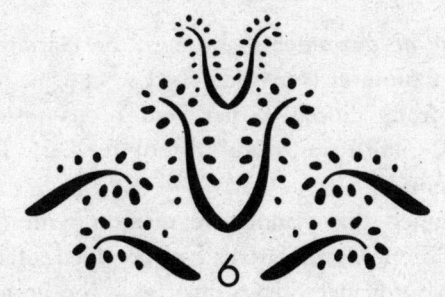

GUERRILLA REWILDING

Three pale moths fluttered through the headlamp beams. Beyond their ethereal flicker a bright five-bar gate, fringed with tufts of spring-green grass and a hawthorn hedge with a little galaxy of white blooms, criss-crossed an expanse of night. On the other side of that gate was more than a cluster of fields soon to become housing. Out there lurked choice.

I switched off the headlights and plunged the moths back into darkness. Blood pulsed in my ears. My heart was beating fast.

I had resolved to never knowingly allow a repeat of what had happened to the Autumn Lady's-tresses on that roadside verge. I had vowed to be the secret saviour of orchids from every development site I could reach. This first mission would be a test of my resolve. I had a small backpack containing plastic bags, a trowel and a head-torch. I had researched the types of orchids that might – just might – be present. Late spring seemed to offer a good chance of finding the Early Purple (*Orchis mascula*). It's among the first orchids to flower in Britain; other species that might be present wouldn't be flowering (and therefore easily visible) for at least another month, by which time diggers could have moved in. I had only ever seen pictures of an Early Purple. The Harraps said it was quite common, but, as experience had shown, statements like that were no guarantee – and even the Harraps conceded that, like the Common Twayblade, the Early Purple had recently disappeared from almost 30 per cent of its historic range. Nevertheless, there I was, ready to look for Early Purples and to rescue them.

I had planned this clandestine mission with (I thought) military precision. Google Maps had provided satellite reconnaissance; an Ordnance Survey map gave the lie of the land, with entry and extraction points; daytime drive-bys allowed me to scope out the site and assess possible risks. In a slow race against time – the turning of the planet and the annual cycle of orchid growth versus the local authority planning-approval process and the as-yet-unconfirmed date diggers would enter those fields – I had settled on the dead of a Saturday night in late April as the moment of reckoning. Any earlier and the first orchids of the year might lack the flower spikes that made them easier to find.

It took facing that gate in the middle of the night for the magnitude of what I was intending to sink in. I had come ready to commit a crime: to trespass and dig out species that were protected, no matter how ineffectually, by the Wildlife and Countryside Act. The latter carried a tariff of six months in prison and a fine of £5,000 per uprooted plant. I would struggle to pay the fine for one plant; caught with half a dozen, I would be bankrupt. A criminal record would probably end my teaching career; a prison sentence could change my life for ever. The law-abiding citizen in me wanted to drive home. I could spend the rest of my life railing against the destruction of native orchids – indeed, orchids, habitats, ecosystems, wildlife everywhere – from a cosy armchair. But, sitting in the car, night pressing against the windscreen, I couldn't get that United Nations phrase out of my head: 'The safeguarding of species is a moral obligation of humankind'. *Safeguarding. Species. Moral. Obligation.* The words became a kind of mantra. I could go home, my future would be secure, but what of the orchids being decimated across the country? Now I had that knowledge, I could not ignore it. *The safeguarding of species.* Who would salvage any last living fragment of those soon-to-be-destroyed fields, if not me? Where orchids were concerned, surely obeying the law in a situation like that was not morally right. *Moral. Obligation.* I took a deep breath, grabbed my thief's bag, opened the door and stepped into the night.

Blackness. An invisible expanse of life. The sigh of distant wind, like waves on an unseen shore. A scattering of stars, oblivion between them. Shrugging the backpack over one shoulder, I switched on the head-torch. Its light glimmered on the steel gate. I clambered over. The cold metal clanged. Vapour from my breath swirled. My heart beat faster. In an unfamiliar medley of guilt, fear and a determined sense of fulfilling human-kind's moral obligation to other species, I jumped off the gate onto thick grass and jogged away, just in case anyone, as unlikely as it was, came to investigate.

Feeling like a character in a horror film, I chased the bouncing torch beam over damp grasses and clover, acutely aware that its reach wasn't far and its fringes were absorbed by night. I kept it directed at ground which tilted away like a leafy sea into the valley. Maps informed me a tree-lined stream flowed there.

I figured the hedgerows around the fields offered the best chance of harbouring orchids – they were probably the oldest, least disturbed areas and were unlikely to have been trampled or over-grazed. I intended to start at the bottom of the field, then race around half its perimeter, scanning the hedges for Early Purples, dig out any I found and make good my getaway. If I didn't find any on one side of the location, I would check the other. The problem was, after what felt like half an hour of walking at pace, my steps loud and rustling in the spring grass, I should have reached the far boundary of the field, but I was still chasing torchlight across what seemed to be an infinite plain now sloping up not down.

I turned around to reorientate. The light petered out into a desert of black air. The glow of the city seeped from one horizon, silhouetting lumpish trees. I heard the muttering wind and the distant speculative whistling of owls. Everything felt unfamiliar. Even the mingled scents of damp grass, roots and earth were alien. Suddenly, nearer than expected, a flock of sheep erupted into a chorus of moans. They sounded enraged, formidable. Primal fears started nibbling at the margins of my mind. Why had I thought wandering around a field in the dead of night looking for a species of orchid I had never encountered in real life was going to succeed? Had I passed into a neighbouring field? Was I going to be stampeded by sheep? Was I going to stumble into the nearby farmhouse and startle a pack of dogs? Where the hell was I?

I decided to keep going until I reached a boundary, then follow it as I had initially planned until it (hopefully) led me back to somewhere I might recognise. The problem was, by night, everything looked different and there were hours until daybreak.

I trotted after the bobbing light – phantom grass, closed daisies, twinkling spiderwebs, silver dew, then, abruptly, a tangled wall. A hedge. Technically a hedge-bank, a long, towering raised bank of earth, common in that corner of England, colonised by a dishevelled mass of vegetation topped with a lattice of elm, hazel and hawthorn. Sweeping the torch across it, I saw ivy, fresh blades of grass, the white filigree of a half-blown dandelion clock, crumbly entrances to rabbit burrows, luminous primroses and the willowy white star clusters of greater stitchwort. Relieved, I turned right and followed its course. Every twenty or so metres the broad buttress of an ivy-embroidered oak rose into obscurity.

By that point I was focused on finding the car, but I had to keep the torch beam half on the hedge-bank to navigate. That was how a strangely upright floral steeple snagged my attention. It was a dark shape, about 40 centimetres tall, rising from the flowing mass of vegetation. I angled the torch. The steeple sharpened into jewels of magenta flecked with pearl. At the top of the steeple the buds were closed mauve knuckles; the lower flowers were open, a centimetre or so across, maybe two centimetres tall, each a deep lilac and indigo with two petals emerging from the top, then drooping like bunny ears to either side of a little hood. Beneath the hood a pearl and citrine cavity receded into a rose spur curving from the rear of each flower. Perhaps that was where nectar was stored. An oval purple lip jutted from the lower half of each flower. There were maybe thirty little flowers on that flower spike which stood like a regal hyacinth on a stout green stem. The stem emerged from a glossy nest of spotted leaves. Each strappy leaf was 20 or 30 centimetres long.

Then I noticed two more steeples rising next to a primrose. My heart gave a squeeze of joy. I could not believe my luck. They were Early Purples, the Male Fool-stones of Gerard's *Herball*, the Early Spotted Orchis of William Curtis's *Flora Londinensis*, about which he remarked, 'In the woods and meadows in most parts of England, no plant more abounds.'[1] How things have changed: in that field, ivy, ferns, grasses, stitchworts and primroses far, far outnumbered the orchids, and soon, if not for me, those three plants would be gone.

I pulled out the trowel. Two of the plants were bigger than I anticipated. The Bees, Marsh Helleborines and Autumn Lady's-tresses had all been fairly compact, but the belt-thick blotchy leaves of those Early Purples spidered across the earth to a diameter of 50 or 60 centimetres. The third plant was about half that size. Maybe the bigger plants were older. Starting with the small one, I commenced my illegal rescue.

The soil was tough, dry clay, packed with roots. It had not rained for a while and I had to angle my whole body weight to drive the blade into the earth, then yank it out and repeat the process, to create a circle around the plant about 10 centimetres from its base. I didn't want to disturb the roots, tuber or leaves. Circle complete, I attempted to work the blade beneath the clump and lever it free. The earth rocked and jerked; the flower spike shook. I worried that, in this struggle for its protection, it might get damaged. Then the clod broke free. It was a heavy ball of dry earth crowned with the orchid rising like a banner. For a moment I was triumphant, satisfied, overjoyed, vindicated, but there was no time to celebrate. I unrolled a plastic bag on the grass, carefully laid the orchid with its earthen boot onto it, then set about digging up the other plants. The faster I could get them into bags and exit, the better.

I struggled. I cursed. I wanted to save them – couldn't they understand that? – but the plants were knitted into the clay. I wormed my bare fingers into the red-brown earth and heaved. Minutes later, breathing hard, I stared, shocked, amazed at the three wild orchids at my feet. I took in their blotched leaves, ragged, quarter-open flowers and spheres of earth, where worms and centipedes were wriggling to safety. I had wrenched a colony of orchids from where they had lived perhaps for decades, and now they lay like cadavers on plastic sheets. I felt like a thief, a moral thief.

I tucked the plastic around the plants and lowered the evidence of my crime into the backpack, nestling them as best I could so they would not crush each other. When I lifted the bag I realised how heavy three orchids in clay could be. My back had healed, but the weight was enough to cause discomfort. If I came across more orchids on the way back to the car, I would just have to return for them another day.

I tracked the hedge-bank up a rise, then turned right, along another hedge. On the other side, a car whooshed past, briefly scattering the night, but that was enough to indicate that beyond was the road I had driven along to reach the field. My getaway vehicle had to be nearby. An eternity later I found the gate, nowhere near where I thought it would be. I clanged over it, opened the car door, placed the incriminating bag in the passenger-seat footwell, started the engine and drove away.

No squad cars screeched into view, sirens wailing, lights pulsing. As I put more miles between me and the crime scene, I began to accept that I had broken the law and rescued orchids. I had broken the law *to* rescue orchids – and I had got away with it. The last nerves and guilt dispersed. My mission had been a success. Just one question: what was I going to do with the rescued orchids?

I had moved into a small terraced house, three rooms upstairs, three downstairs, in the middle of a row of identical houses, in an area of the city full of such houses, where my only outdoor space was a small paved backyard bounded by tall brick walls. I moved the rescued Early Purples to pots there, using sand and grit to fill in the space around the clay still clinging to them, but very soon they were not the only rescued orchids in my care. As the numbers increased, the question of what to do with them became more pressing.

I had abandoned problematic midnight excursions in favour of raids at first light. Confidence supplanted nerves; triumph and determination replaced guilt. By August, in unassuming fields, sometimes in the company of ruminating cows and edgy sheep, I had carried out nearly a dozen operations. Less than half of them were successful, but when I did find orchids, there was always more than one plant. Orchids, it seemed, like the company of other orchids.

I had returned to that first site and found six more Early Purples, which I dug out and carried to safety. I went back again in late May and found nothing, but around the same time, at a different location on the city outskirts, where a large advertising board declared the imminent arrival of 'The Elms, A New Luxury Development' alongside a giant picture (oh, the irony) of two children frolicking through a wild-flower meadow, I wormed my way past the steel fencing and encountered – and saved – my first Common Spotted Orchids (*Dachtylorhiza fuchsii*). They beckoned from the damp, shady corner of a field like a very small forest of pale lilac, black-and-white-speckled spires hovering above grass and buttercups, oblivious to the fact that they were in danger of being entombed in concrete. I saved a dozen plants all told.

In early June, after failing to find them on that cliff-top Site of Special Scientific Interest a year earlier, I came across five Common Twayblades on a threadbare rise under a lonely beech tree on a site destined to become a retail park. Their twin big oval leaves seemed to reach like two cupped hands from the earth; from the centre of the leaves rose a single tall, green column decked with flowers which looked less like Gerard's gnats and goslings and more like miniature people blown from lime-green glass by El Greco, their long legs dangling, their faces with tiny puckered lips, little arms reaching out in welcome and gratitude. They were enlaced with silver spiderwebs and were just as enchanting as I had dared imagine. I dug them out with reverence; they could have been living there for decades, each plant waiting twenty years to flower.

Also in June, near an abandoned car tyre at a marshy site approved to become an extension of an industrial estate, I found an array of Southern Marsh Orchids (*Dactylorhiza praetermissa*), their bright streaks like brash mauve Mediterranean firebrands, as if a renegade artist had mixed alizarin crimson and French vermilion on a palette knife and smeared the result onto a modest sap-green canvas.

All Marsh Orchids currently fall under the genus *Dactylorhiza*, which translates as 'finger roots', an association inherited from early herbalists like William Turner, who, in providing the first account of this genus in English, describes their roots as 'like a mannes hand'; *praetermissa* means 'overlooked' because, although Marsh Orchids have been written about since the sixteenth century, Southern Marsh Orchids weren't recognised as a separate species until 1914, when English botanist George Claridge Druce proved that they were.[2]

According to the environmental impact statement, the development intended to preserve and improve that boggy patch as a wildlife sanctuary. I had doubts about how effective that plan might be, so I hedged the orchids' bets and took a handful, leaving the rest to their fate beside the tyre.

So it was that come the close of that summer I had more than three dozen refugee plants of four species in my yard. By then most of their leaves had browned and their flowers withered as they bowed out for that year. But while those species retreated, I knew that Autumn Lady's-tresses' flowering season was weeks away and my yard was already getting full. If I rescued a dozen Autumn Lady's-tresses, where would I put them?

Space was not the only issue. Was I going to let all those exiles live out their lives in a yard? Forty or fifty rescued plants were barely a drop in the ocean of what needed to be done to save the nation's orchids. As long as they stayed in that yard they were safe, but they were not boosting the country's orchid population.

My expeditions indicated that urban expansion was killing orchids. Maybe the numbers per development were small, but given the huge amount of development across the country and the lack of monitoring of the cumulative effect on the nation's wildlife, thousands of rare plants were probably succumbing to concrete every year.[3] As long as there were developments on pastureland, woodland and parkland, orchids would need rescuing, and I seemed to be alone in heeding the moral obligation to do anything about it.

The problem was, I couldn't keep doing it in that way. I had a full-time job, and I was still attempting to visit sites around the country to see other native orchids for my own edification. Applications for new developments kept flooding into the local planning authority, leaving me hundreds of pages to trawl through. Very early Sunday-morning rescue missions needed to be planned, then executed, and the pots in the yard had to be shuffled around so their residents received the right balance of sun, warmth and water. It was all too much of a draw on my time. I needed a new plan, some way to rescue orchids without their numbers being constrained by those four walls.

That was when I started thinking about rewilding with orchids, but not just 'normal' rewilding: *guerrilla* orchid rewilding.

'Rewilding' did not exist when I was young. At that time there were around three billion – three *billion* – fewer people on the planet and around 60 per cent of the world's wild creatures had not disappeared.[4] As a response to humanity's relentless assault on nature, the concept of rewilding first appeared in print in the mid-1990s. The brainchild of radical conservationists, it was a way of giving nature a hand to move back into areas where it had been evicted or degraded by human activity (or inactivity).[5]

Although initially a radical concept, rewilding has become a widely accepted approach which offers benefits for carbon dioxide sequestration and countering biodiversity and habitat loss. My dad's mini-meadow is an example of very small-scale rewilding. On a grander level, there are accounts of rewilded farms, the creation of woods and wild-flower meadows and the reintroduction of beavers into waterways so they can recreate ecosystems that have been lost for centuries.[6] Many rewilders advocate reintroducing extinct species – wolves, lynx, Eurasian bison, or other species that can carry out important roles once performed by large mammals long lost from Britain.[7] However, rewilding does not always require species reintroduction; once established, like Dad's meadow, those spaces can attract lost species, including orchids, all on their own.

The concept of taking little pieces of Britain and effectively turning back time in them to an era when intact ecosystems flourished was an attractive, probably incurably romantic idea. If I had a field to put the refugee orchids in, it would offer a real opportunity to stall the decline of orchids. The problem was I had no field. I didn't even have a garden.

It occurred to me that I could secretly rehome the orchid refugees in Dad's mini-meadow, but I didn't want to face awkward questions about where those new plants were coming from. I could take them to some of those isolated, protected places like the dune slacks I had visited the previous year – perhaps to national parks or Natural England-designated Sites of Special Scientific Interest. But I knew from experience (for example, a trip to a field carpeted with thousands of Green-winged Orchids in the middle of nowhere in Somerset) that those areas tended to be reachable only by private vehicle. Their remoteness explained why the orchids had been spared extinction, but reaching them would involve a fair bit of mileage, petrol, time and effort (including the effort of finding out where the places are). On top of that, some required special permission to enter.

While transplanting the refugee orchids to those places would be a safe option, it struck me that Britain's orchids, with their intricate beauty and all their rich, largely forgotten cultural history, did not *need* to be the preserve of determined weekend naturalists on pilgrimages to isolated reserves – those lost roadside Autumn Lady's-tresses proved that.

In the course of my orchid research I had come across a searchable map curated by the Botanical Society of Britain and Ireland. It was regularly updated with information from thousands of volunteers with the location of various species, including orchids. Filtering the database by decade or year presented a picture, sometimes stretching back to the eighteenth century, that showed, through dwindling numbers of coloured dots, each representing a small area containing orchids, an accelerating decrease in orchid sightings.

By 2011 for some species many of the dots were concentrated on those isolated reserves, but the database also offered reports of unexpected orchid sightings on roundabouts, next to bus stations and sites such as 'beside the footbridge between Tesco and the industrial park'. A Google search of media reports concerning unexpected orchid appearances showed that when native orchids popped up in everyday places they often generated excitement, even pride, in the community; conversely, when the local council or developers 'accidentally' destroyed them, communities reacted with outrage.[8]

Those unexpected occurrences interested me. They mirrored my first encounter with that Bee Orchid in my parents' garden and the Autumn Lady's-tresses on the verge by the main road. It hit me that I could rewild with the orchid refugees in a way that engineered that kind of unexpected orchid encounter. The more I thought about it, the more it seemed I had solved the issue of the backyard orchids.

I wondered where I could plant them. Was a park public land? A roadside verge? A motorway siding? Then it occurred to me that those questions were immaterial. I was a British citizen; Britain was my land. Yes, vast areas had been claimed by others as private property, and entering them might (technically) be trespass, but my ancestors' bones lay in that earth. They had fought and died for it; farmed and shaped it. Surely that counted for something, and if all it meant was that I had a claim to a few square metres of earth for the purpose of rehoming endangered native species, so be it.

Considering all of Britain as 'mine' to rewild with rescued orchids allowed those refugee plants to become (in my mind at least) a bit like a guerrilla army raising the profile of Britain's neglected natives. Orchids unexpectedly appearing in villages, towns and cities might inspire people to take more interest in them, and, just as they had with me, might encourage people to do more to save them – and maybe not only orchids, but other threatened elements of Britain's natural world too.

So it was that, having saved them from destruction, I decided to bring a few of Britain's most captivating flowers back to the people.

It turned out that secretly planting urban areas with orchids is more complicated than digging a few holes in the local park and dropping plants into them. In most instances, introducing plants to the landscape, even native plants, while not illegal, contravenes the codes of practice of several influential organisations. The code of conduct adopted by the Wild Flower

Society, written by members of the Botanical Society of Britain and Ireland (BSBI) and promoted by the charity Plantlife, says this about introducing plants (and seed) to the wild:

> The main emphasis of conservation is to maintain native plants within their natural ranges. Introductions may disturb natural patterns of distribution, which can be subtle and involve sub-species and varieties. Many plants have been introduced into the wrong places, and inappropriate, even foreign, strains have been released. There is therefore a strong presumption against casual introductions. Do not introduce seed or other living plant material to the wild unless this is part of a well organised scheme sanctioned by your local wildlife trust or botanical society, or by one of the statutory conservation organisations.[9]

I understood the purpose of this guidance. All across the world, humans have taken species out of one environment and dropped them into another, where they have become plagues. Britain plays host to around 2,000 invasive non-native marine, terrestrial and freshwater species, 10–15 per cent of which are considered to be detrimental to biodiversity, economy or society.[10] Among these, plants such as Japanese knotweed, Himalayan balsam and Canadian pondweed have overrun native flora and had a disastrous impact on ecosystems.

Overtly detrimental effects aside, having unexpected plants appearing where they 'shouldn't' can play havoc with genetics and with research into natural distribution and the accurate monitoring of our wildlife's disappearance. Undoubtedly, this is why scientists observing species are cautious about recording newly appearing species (have they appeared by natural dispersal or not?) and follow strict guidelines regarding 'wild' populations that have been artificially 'bolstered' or created, even for conservation purposes.[11] This struck me as understandable, but also a little strange – a threatened species growing in the wild, whether put there by a person or growing from a windblown seed, was still, well, a threatened species growing in the 'wild'.

I found myself questioning the code of conduct. All studies

demonstrated that this timid attitude towards conservation was not working. For decades, species had been disappearing at accelerating rates. No big changes to conservation methodology were being announced; spreading seeds and planting plants in the wild continued to be discouraged. But would an urban space fall under the code's definition of 'natural' or 'wild'? I doubted it. What about that phrase 'introduced into the wrong places'? A British native orchid is . . . *native*, so how would it be 'wrong' to reintroduce it to its native land? Did the code actually make sense, given that plants have evolved to naturally disperse wherever they can, including colonising new areas?[12]

The code seemed to be suggesting that, unless properly authorised and regulated by other humans, people should not assist nature in what it had evolved to do over hundreds of millions of years. Why, then, did the code not support the rigorous weeding-out of plants that had appeared in the 'wrong places'? It was also noticeably quiet about all the non-native crops (potato, corn, tomato) growing across hundreds of square miles of the country. And what about genetically modified crops? And the non-native pine species planted across England? Or the plants imported into the country, worth billions of pounds a year to the horticultural industry, which can cross-pollinate with native species and/or become naturalised?

Even the idea of 'native species' is complex: the chestnut (*Castanea sativa*) and sycamore (*Acer pseudoplatanus*) are non-native trees in England, and many 'native' wild plants are introductions. The Normans introduced betony (*Stachys officinalis*); returning Crusaders brought comfrey (*Symphytum officinale*). These medicinal herbs spread out of physic gardens into the wild.[13] Taking all that into account, it seemed that Britain was already a melting pot of introduced flora (not to mention fauna: rabbits, sheep, fallow deer, pheasants . . .).[14]

There were also practical issues to consider. Guerrilla-planting orchids cannot happen just anywhere, any time – at least not without compromising the transplanted orchid's chances of survival. All plants have adapted to particular conditions: you don't keep a cactus wet and dark (unless you want it to die) any more than you would attempt to grow an oak tree in a desert. Roses do best in acidic soil; other plants like neutral or alkaline conditions. As for native orchids, most prefer slightly alkaline earth (which explains why dune slacks are a favourite habitat: the sand is ground-down shells, rich in calcium and therefore alkaline).[15]

I purchased a digital soil pH meter for a few pounds on eBay and figured it would assist in choosing possible sites to plant. I spent some time strolling around the neighbourhood, loitering near verges, checking no one was watching, then pulling out the pH meter, plunging its probe into the earth, taking a reading and casually walking on. The neighbourhood proved to be mostly neutral or slightly acidic, which was fine for the local species I had to rehome. However, pH was far from the only consideration.

Orchids detest synthetic chemicals, so I would have to avoid places likely to be fertilised or sprayed with herbicides (I figured regular spraying with dog pee would be enough to change the chemical composition of the earth too, so I also bypassed areas popular with dog walkers). Ploughing and digging earth kills orchids, so do strimmers. That further narrowed possible planting locations. Many species (Marsh Orchids excepted) prefer well-drained conditions; I needed to consider that too. There were also the right light levels to find. Common Twayblade and Common Spotted are reasonably happy in relative shade; Southern Marsh and Bee prefer brighter conditions; Early Purple seem not to mind either way. Other species that I was not engaged with at that time require even more specific habitats, such as the edges of bogs or gloomy areas of woodland near beech trees.

The fact that slow-growing orchids cannot compete with faster-growing species also had to be taken into account. The guerrilla orchids needed a location that was sparsely vegetated or where other plants did not grow tall. On top of all that, wherever I chose to plant them, my orchid 'influencers' had to be visible. People needed to encounter them. In this respect, I was fortunate that the orchids at my disposal, although not the grandest, rarest or most spectacular (like Military, Lady, Lizard), were reasonably conspicuous, unlike the small yet exquisite Musk, Small White, Dense-flowered and Lesser Twayblade that can be found (if you are lucky) in a few parts of the country.

Finally, I had to deploy my troops inconspicuously. The orchids needed to appear as a surprise, not as the work of a guy with a trowel dispatching plants from a bag. To do that, I would have to venture out before dawn. I'd need to dispense with bulky terracotta pots. That meant delicately digging those tubers and rhizomes out of the earth that had nurtured them for years before transporting them to their new home. Orchids tend to be happiest for that to take place at the time of year when the plant is dormant. Fortunately, as autumn approached, the refugees in my care were turning to dormancy.

Dormant orchid rhizomes and tubers can be handled like the bulbs of daffodils and tulips (although, unlike those bulbs, orchid tubers and rhizomes must be kept slightly moist, in damp sand or moss). Transporting tubers like that, rather than in bulky pots of sand, clay and grit, is much more convenient for a guerrilla orchid rewilder, but it required their careful extraction from the pots in the yard. That was when I first got a good look at those orchid roots that for so many centuries had interested herbalists.

The first tuber I unearthed – a Common Spotted, its leaves and flowers withered and fallen away – was a compact root system resembling a small figure with pale, tapering limbs a few centimetres long stretching out from a thicker, vaguely oblong torso-shaped bulb, also a couple of centimetres long, about a centimetre thick and the colour of vellum. This was the root Gerard and others called the new 'full or fat' replacement tuber. At the top of that little torso, by way of a head, emerged a small white nub. That was the growing point from which the next year's growth would come. Tangled up with its 'limbs' was a shrunken echo of the same structure: the dying roots of that year's plant, the old tuber ('drie or barren . . . withered or shrivelled').[16] Hence, at certain times of year, orchids have two twinned 'roots'. According to ancient belief, the plump new one acts as an aphrodisiac, while the withered, spent one has the opposite effect; one produces male children, the other female.

That is how many of Britain's orchids grow: each year nutrients created via the leaves go into the formation of that new twin tuber which stores energy for the subsequent year. In the meantime, energy from that year's tuber produces leaves, flowers and seeds. Once exhausted, at the end of the season, that year's tuber dies and the new one waits to spring into life.[17] Some species are dormant during the winter and have leaves in the spring or summer ('summer green'); others produce flowers and seed earlier in the year, then go dormant for the summer and produce leaves for the next season during the autumn and winter ('winter green').

I found the Southern Marsh Orchid rhizomes were similar to the Common Spotted – perhaps unsurprising, as they both belong to the *Dactylorhiza* genus – but the Common Twayblade pots revealed structures more like a small sat-upon bird's nest, as suggested by the name of its genus, *Neottia* (from the Greek for 'nest'). Of the species in my care, only the Early Purples possessed those infamous small, brown-skinned, vaguely testicle-shaped tubers from which orchids get their name.[18]

That was my first lesson in the underground world of native orchids. Some have spherical tubers (including the Bee, Early Purple, Military, Monkey); others have tubers like small parsnips, long sweet potatoes or carrots (Autumn Lady's-tresses and the Greater and Lesser Butterflies). Some have hand- or witchy-figure-like tubers (Common Spotted and other Marsh Orchids), while others (including Common Twayblade and Lady's Slipper) grow storage rhizomes like a mess of tangled twigs.

Whether bird's nests, witchy figures or little balls, I carefully folded the first dozen dormant orchids into tissue-paper cones, each containing a spoonful of the earth from which they had been rescued mixed with a little grit and sand. This would be their survival capsule to aid them in their rehoming in the city.

The next morning, before sunrise, I placed individual refugee packets into my pockets, put my trusty trowel into a supermarket bag and stepped out to begin urban guerrilla orchid rewilding.

Five o'clock in the morning, a Sunday, early October. My destination – the orchids' destination – was a park occupied partly by a large playground, partly by a turf-covered bank where the grass grew short. I felt exposed without the hedgerows which had concealed my activities around development sites at the edge of the city where birdsong had serenaded the rising summer sun. Those October streets lacked privacy.

Glazed with lamplight, the concrete and tarmac world was surprisingly busy. Ragged groups of students staggered home talking loudly about parties. Souped-up hatchbacks roared down the main arteries. Hungry revellers queued outside kebab shops. I kept my head down and my pace up, until I turned off the main thoroughfares into a darker side street and came to the entrance to the park.

Beyond the black iron railings was a large grey lawn. Buttresses of linden and chestnut reached for hazy clouds tinged orange by the city's glow. The lawn, the benches, the shadows beneath the trees, the concrete paths, the playground – all were deserted. I entered through a pair of ornate iron gates and proceeded down a wide path. Out of the corner of my eye the white tip of a fox's tail bobbed, luminous, as it bounded away.

A red metal fence enclosed the playground. Opening the magnetic gate, I passed the toddler swings and seesaw. They were eerily still. I knelt at a spot near the farthest fence that was sufficiently out of the way of little running feet to avoid being trampled, yet still visible enough for an observant mum, dad or youngster to notice something beautiful.

I unwrapped the trowel, took out my torch and laid it on the grass, shielding its light with my body. I stabbed the trowel into the earth, levering it back and forth until it formed a slit a few centimetres deep. I removed one of the paper parcels from my pocket, opened it, sprinkled a little of the sand mixture into the slit, gently removed a Common Spotted Orchid rhizome and eased it in, growing point upwards. After persuading the longer limb-like parts of the tuber to go in too, I sprinkled the rest of the earth, sand and grit on top, then pushed the slit shut. Once closed, there was no sign that it had ever been open. I repeated the process a foot away: stab, open, sprinkle, drop, sprinkle, close. I did the same twice more. By planting those orchids in small groups, I was hoping they might be cross-pollinated by insects, then form seeds and create a self-sustaining colony. That was the hope. Time would tell.

I took up the torch and trowel, scanned the park – no witnesses – exited the playground and repeated the process on the short turf bank nearby. Ten minutes later, I left the park via the same iron gates. There was no visible sign of what I had done, but to me the world had changed. The change was tiny, but knowing that it involved saving orchids that would otherwise have died, and that I, like a covert operative on the side of nature, had effected that change, made each damp gust of that blustery morning a flurry of hope and promise.

ORCHID ORCHARDS

It would take a year or so to know if my rewilding attempt had been successful and several more, maybe a decade, to know whether any of those colonies survived long enough to become self-sustaining. Despite that, I was pretty confident that the future lay in one-man guerrilla orchid rewilding. So over the course of that October, I crept out before sunrise, in the company of foxes, drunks and boy racers, to guerrilla-plant all my rescued orchids in various places around the neighbourhood – churchyards, other parks, the edges of playing fields.

After a handful of trips I ran out of orchids to rehome. A small part of that small city had easily swallowed all of the plants I had spent months rescuing. If every one of those few dozen orchids flourished, they might add specks of inspiration to part of the city, but that city had many parts. It would take hundreds of native orchids to raise their profile. Getting that number of plants depended on me saving enough of them, and that, not to mention my aspiration to guerrilla-rewild England, would take thousands of plants.

I was facing a paradox: to relocate thousands of plants required *more* land to be developed and *more* habitat to be destroyed simply so that I could rescue and rewild orchids. That made no sense. Besides, for someone with a full-time job (I had been too swamped with work to search for Autumn Lady's-tresses that year), it was a daunting task. I would have to scale back my ambitions or become more efficient.

About half of my visits to development sites revealed no orchids, and while it was profoundly rewarding to rescue those I could, at best I was only very slightly slowing the decline of those few species growing nearby. The process of researching sites and rescuing and rewilding a few dozen plants from them had to be streamlined. During the subsequent year I tried to do this by using binoculars to scan the sites instead of walking all around them. It wasn't a satisfactory solution, and I usually ended up tramping around them anyway, but over several summer dawns, I managed to sneak through enough perimeter fences and gates hung with warning signs to rescue a couple of dozen or so more plants without getting caught.

Then a new problem emerged. I noticed that many of the first generation of guerrilla orchids had failed to surface. This was deeply dispiriting. It was time, effort and, most of all, hope wasted. Sometimes it was because areas that had looked sparsely vegetated in the autumn when I dropped tubers into them erupted into long, tangled growth the next spring. That long grass swamped the orchids. In other areas it was more difficult to pin down a cause. Had it been too dry when I planted them? Too wet? Too late or early in the season? Had they succumbed to disease? Slugs? Rats? Was the chemical composition of the soil wrong? I wasn't sure what to do. The whole premise of guerrilla orchids was to make native orchids visible in public spaces, but I had no control of those spaces. Sure, guerrilla orchids would have a better chance of survival in a nature reserve, but those are not the places 'everyone' passes.

With few alternatives coming to mind, I persevered with my plan, seeking new spaces which might offer better refuge to the rescued plants: a bank beside a main road, the grounds of where I worked, inside a ruined church. Nevertheless, the prospects for my urban orchids were not looking great. Then one day I read an article about the Sainsbury Orchid Conservation Project at the Royal Botanic Gardens at Kew. According to the article, Kew, an internationally recognised institution staffed with orchid experts (who would probably not have approved in the slightest of my well-intentioned bumbling), had successfully reintroduced wild orchids, both in the grounds of the botanic gardens and in some protected sites around the country (those places requiring a dedicated trip to visit).

Although orchids in out-of-the-way places did not match my urban orchid-rewilding ethos, Kew's orchid reintroductions got me thinking. They had collected orchid seed and propagated, then planted the young plants. Therein lay the answer. To *really* boost orchid numbers required more than rescuing them: I needed to propagate them. I needed native orchid seed.

Generally speaking, there is only one way to get orchid seed: collect it yourself. No commercial seed supplier sells it. Why not? Stick an acorn, sunflower seed, carrot seed, seeds of chilli, cress or cornflower into earth with the right amount of heat, moisture and light and, after some time, hey presto: roots, stems, leaves, flowers. Not so with orchids. At least, not in the same way.

If someone buys a commercially produced pack of orchid seed (which, usually, they can't) and empties it into a pot or flowerbed, if germination occurs at all, it will take years before that plant flowers, by which time it will probably have been dug over, fertilised, mulched and killed. For those reasons, large commercial producers hardly ever offer native orchid plants for sale. Growing them from seed to mature plant takes quite a long time; unless you make a little plant expensive, profit margins are feeble.

Without the option of buying orchid seed, obtaining it in person gave me two options: collecting seed from wild plants (which, if it involved Schedule 8 species without the correct permissions, would put me in conflict with the law), or creating an 'orchid orchard' from orchids in my care (which, given that

I had obtained the plants by rescuing them without permission, meant I would already be in conflict with the law). Could I have asked for permission to collect seed? Sure. Was I likely to get it? Doubtful. I was unknown. Unqualified. Whoever had the authority to grant me permission would have considered me highly suspect. Besides, what little spare time I had I wanted to spend saving orchids, not writing emails, and if the relevant person said no, was that going to stop me from rescuing orchids, getting seed or building my orchid orchard?

An 'orchid orchard' is the name for a few orchids kept and pollinated, often in a controlled environment, so their seed can be gathered for micropropagation in a lab. Done correctly, this produces lots of baby orchids. With luck and the right conditions, these become lots of adult orchids. This technique has been used successfully in large commercial ventures involving tropical orchids and the houseplant market. It is also one of the methods Kew's experts use.

It seemed to me that orchid orchards didn't have to be the preserve of large operations, and seed didn't have to go to a laboratory. After all, for tens of millions of years seed evolved to germinate just fine in a non-sterile world. I imagined gathering orchid seed and scattering it across the land, producing hundreds of plants (at least, so I thought at the time). This would reduce the need to trawl the ecological reports of proposed developments and trespass on development sites in the hope of stumbling across orchids. But there was a problem. I was orchid orchard-less. I had already deployed all 'my' rescued orchids into the urban battleground.

I resolved to turn the next plants I rescued into my orchard. Then it occurred to me that I already (sort of) had one: the Bees in Dad's mini-meadow. There were also the orchids I had already planted around the neighbourhood. I simply needed to wait for them to flower, get pollinated, then collect their seed, scatter it, grow it and . . . I didn't yet know much about orchid seed let alone its germination, so my ideas were vague. However, as seed appeared to be central to the optimised guerrilla-orchid venture, I figured the next step was to learn how to pollinate an orchid. If I did, I could guarantee seed. If I guaranteed seed, I could scatter it. That would save the effort of creeping out at dawn to dig holes for tubers.

As a youngster I had watched my mum in the greenhouse pollinating tomato plants. She used a small paintbrush, the bristles scuzzy with use. One day, as she tickled the brush-tip, thick and powdery with grains of yellow tomato pollen, across dozens of bright yellow tomato flowers, she explained that she was being a bee. She handed me the brush; I copied her. Weeks later, the hand-pollinated plants in the greenhouse were laden with fat, sweet red fruit. I guessed orchids worked in much the same way. I was wrong.

Orchid flowers are not just different from tomato flowers. They are different from every other family of flowering plant. To appreciate just how different, it's worth taking a look at an average flower.

Flowers are the reproductive organs of plants. For simplicity's sake, let's say that most non-orchid flowers are circular with four basic components. Tucked in under the *petals* are some *sepals*. Sepals are tougher than petals and create a protective covering around the flower when it's in bud. When they open, the flower unfurls and the sepals usually curl out of sight beneath the open flower. For most people the petals are the attractive part, colourful, rounded or pointed, curved, radiate, cupped. Their attraction to humans is largely incidental: for the plant their main purpose is to act as attractive landing pads for insect pollinators.

Enticed by scent, colour and the promise of nectar rewards, an insect landing on the petals crawls towards the centre of the flower. Approaching the centre, the insect crosses the male parts of the flower, the *stamen*, which includes *anthers*. Anthers are filaments topped with blobs of dusty pollen grain. In the very centre of the flower, surrounded by the anthers, stands the female structure, the *pistil*. The tip of the pistil is usually flattish. This surface is the *stigma*. The column supporting the stigma is the *style*. Nestled within the base of the style, deep inside the flower, sits the *ovary* (like the plant's womb). Inside that, the *ovule* or *ovules* (the flower's eggs) wait to be fertilised.

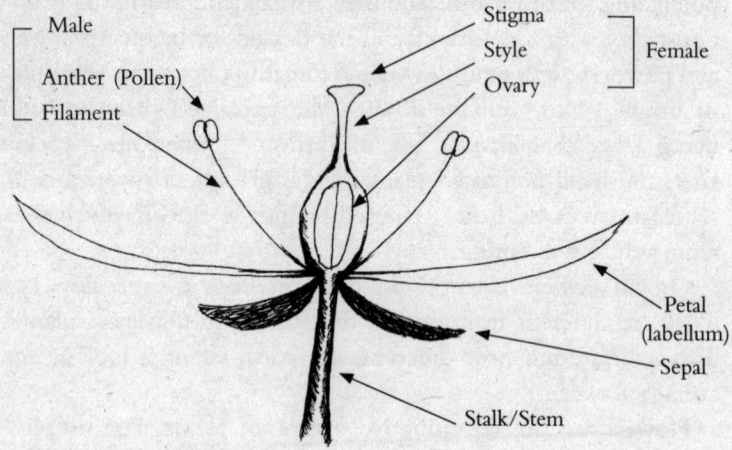

Cross-section of a typical non-orchid flower.

Each pollen grain is a tiny armoured capsule designed to make an important journey. A microscope reveals that they are intricately patterned, often spiked, with everything designed to stick to a pollinating insect. Inside each capsule sits the flower's precious male *gametophytes* (in essence, the plant's sperm). As the insect moves across the stamen, pollen grains stick to its carapace and hairy legs (or to the bristles of Mum's old paintbrush) and detach from the anther. Now recruited to carry one or more of those minute armoured capsules, if the insect crawls across a compatible stigma, some of them may stick to the stigma and detach from the insect.

The stigma produces secretions to encourage compatible pollen to sprout a pollen tube. This tube grows down into the stigma, through the style, carrying with it two sperm cells. Guided by hormones, the tip of the pollen tube grows into the ovary and reaches the ovule. A double fertilisation occurs. One sperm cell fertilises the egg. This becomes the plant embryo, ultimately growing into the plant. The other sperm cell fuses with the ovule's two nuclei. This fertilises the *endosperm*. A bit like the way yolk becomes a food store for baby birds developing inside eggs, the endosperm becomes the food supply for the growing plant embryo. In terms of Mum's pollinated tomato flowers, the red tomato fruit is neither embryo nor nuclei but the swollen ovary, its juicy pulp containing the tomato embryos inside their seed cases together with their endosperm food supplies.

So far, so good? That is how most plant embryos together with their food stores are fertilised: the embryo becomes the plant, the endosperm the packed lunch feeding it until it pushes through the earth, extends its first green leaves and becomes self-sufficient (via photosynthesis).

Now the bad news: orchids do not work like that.

8

CURIOUS CONTRIVANCES

As interest in orchids shifted from their roots to their flowers and seeds, Europe's botanists, herbalists and apothecaries noticed that orchids didn't follow the same rules as other flowering plants. This presented them with a number of conundrums. Why were the flowers, pollen, stigmas and anthers of orchids so different? Why did they have such bizarre shapes? Where did these plants come from?

In 1539, the priest, physician and botanist Hieronymus 'Tragus' Bock offered his learned opinion: orchids grew where birds' semen spilled on the ground.[1] Athanasius Kircher expanded on this idea, observing that orchids grew spontaneously from 'the latent survival force in the cadavers of certain animals [and where] animal semen falls to the ground in mountains and meadows'.[2] Others weren't convinced, and the debate continued well into the nineteenth century when, although microscopes revealed far more detail, they also raised questions: Where exactly was the pollen? Was that dust left over after a flower faded *really* viable seed? Why did these flowers look so different from others?

Obviously, native orchids had played a central role in the early years of scientific botanical study, but to kick-start my orchid orchard into production I needed answers and a practical guide. I needed to leave the debates of eighteenth-century botanists behind. The book I really needed, the one which finally managed to explain how these strange flowers worked, was published at the height of Victorian Orchid Mania and was the work of perhaps Britain's greatest scientist.

During much of 1860 and early 1861, while arguments ignited by the groundbreaking book *On the Origin of Species* rumbled around the country, its author, Charles Darwin, was busy at home in Kent boiling the flesh off dead hens, ducks, pigeons and rabbits, so that he could measure their bones for a study on the effects of cross-breeding on domestic animals. Perhaps it's unsurprising that at this time orchids, Darwin's sideline interest for several years, became an increasingly appealing diversion. By the summer of 1861, the stewed carcasses were set aside and orchids became Darwin's latest 'passion'.[3]

A year earlier, Darwin's eldest daughter, Henrietta ('Etty'), had suffered a bad case of typhus. By mid-1861, she was yet to fully recover. To help her, the entire Darwin family – Charles, his wife, Emma, six of their children (aged ten to seventeen) and eight more 'souls' (servants, governess, nurses, maids), plus '¾ of a tun of luggage' – travelled from Kent to take in the revitalising sea air of the fishing village-cum-health spa of Torquay in south Devon.[4] There, almost exactly a century and a half before I unsuccessfully tried to find Common Twayblades in the area, the Darwins spent eight weeks at 2 Hesketh Crescent, an imposing Regency building with panoramic views over the sea.

Judging from the flurry of correspondence and reports of hours spent studying orchids, dissecting and writing, observing local Bee Orchids forming seed capsules, counting how many flowers were pollinated in a colony of Pyramidals and watching how 'humble-bees' visited hundreds of nearby Autumn Lady's-tresses, it seems that a large part of Darwin's time in Torquay was less centred on family matters than on investigating how orchids got pollinated.

Years earlier, Darwin had met the Scottish botanist Robert Brown, who in 1831 had published an influential paper on the structure of orchid flowers. Brown encouraged Darwin to read a somewhat controversial book by the German botanist Christian Konrad Sprengel.[5] Published in 1793, Sprengel's work put forward the novel idea that insects pollinated plants. He even went so far as to suggest that some plants, particularly orchids, deceived insects into visiting them by *pretending* to produce nectar. Darwin disagreed with aspects of these studies but he used them as springboards for his investigations. He anticipated his research would lead to an interesting but not very important paper.[6] In the end, it became a book which revolutionised the understanding of orchids.

The book's first draft was completed in the second half of 1861. This didn't give Darwin enough time to do all his own research. To assist him, he published letters in the *Gardeners' Chronicle* and the *Entomologist's Weekly Intelligencer* asking readers to get in touch about orchid pollination and pollinators.[7] Entomologists sent Darwin insects they had caught with orchid pollen stuck to them. As for orchids, he requested that specimens be sent to 2 Hesketh Crescent in a 'cylindrical old cocoa or snuff cannister' kept moist with a little damp moss.[8]

When he discovered that no Marsh Helleborines grew in the area, Darwin tapped up a correspondent in the Isle of Wight ('Should you think me very unreasonable if I were to ask you to send me a few . . .?'), before asking for more ten days later ('if you are not utterly sick and weary of me and my requests').[9] Sir Charles Lyell, a geologist and pioneer explorer of the effects of climate change, sent Darwin forty-nine Marsh Helleborine flower spikes. Darwin's friend and director of the Royal Botanic Gardens at Kew, Joseph Hooker, and the renowned orchid importers, Messrs Veitch and Sons, sent him tropical specimens. Bingham Malden, vicar of Sheldwich in Kent, sent boxes including some of the rarest orchids in the land – Lizard, Military and Lady. Meanwhile, George Chichester Oxenden, second son of a baronet and author of satirical verse, supplied him with multiple orchids, including the Burnt Tip. Desperate to supply Darwin with a fresh sample of the rare Coralroot Orchid but unable to find any nearby, one contact tried to persuade a professor at Edinburgh University to send a student into the nearby Ravelrig Bog to find one.[10] It seems this request went unanswered.

Darwin also relied on his sons. In May he had written to his eldest, William, at that time in Wales, requesting that 'If you should find any rare orchids send me some in bud & open'.[11] Meanwhile, sixteen-year-old George spent much of that damp summer on the cliff tops around Torquay deploying bell jars (to place over orchids to see if they really *were* pollinated by flying insects) and nets (to catch insects to see if any orchid pollen was stuck to them). George stayed out late at night netting moths near Pyramidal Orchids to see exactly where the orchid pollen stuck to them and spent hours uncovering how the tiny Musk Orchid was pollinated by very small flies.[12]

All this research poured into Darwin's book, *On the Various Contrivances by which British and Foreign Orchids are Fertilised by Insects, and On the Good Effects of Intercrossing*, published in spring 1862 by John Murray (who insisted on adding 'and Foreign' to the title, probably to appeal to the botanical vogue of the day). As the first book Darwin published after *On the Origin of Species* (1859), it tends to be overlooked, but Darwin's orchid book is perhaps equally groundbreaking. Combined with exquisite woodcuts by George Sowerby, Darwin's fascinating text provides the perfect how-to guide for pollinating Britain's orchids.

Obviously, Darwin didn't write *On the Various Contrivances* to aid a guerrilla orchid rewilder a hundred and fifty years after publication. Darwin actually intended it to provide the hard scientific support for his theory of natural selection, support he had, for various reasons, not included in *On the Origin of Species*.

In my volume 'On the Origin of Species' I gave only general reasons for the belief that it is an almost universal law of nature that the higher organic beings require an occasional cross with another individual . . . Having been blamed for propounding this doctrine without giving ample facts, for which I had not sufficient space in that work, I wish here to show that I have not spoken without having gone into details.[13]

Darwin found that orchids possessed an array of modifications 'almost as perfect as any of the most beautiful adaptations in the animal kingdom', which he called 'curious contrivances'.[14]

Published at the height of the Victorian obsession with tropical orchids and written by a man who, by that point, possessed a sizeable collection of them, Darwin's book bucked the popular trend by focusing on British species.[15] A field near the Darwins' house where he and his family went for picnics was home to many native species. Charles had observed them for many years and over that period several native species made their way into the Darwins' garden. These were not natural introductions: Charles admitted to digging up native orchids and transplanting them.[16]

On the Various Contrivances discusses Early Purple, Pyramidal, Fly, Spider, Bee, Frog, Fragrant, Greater and Lesser Butterfly, Marsh and Sword-leaved Helleborines, Creeping Lady's-tresses, Autumn Lady's-tresses, Bog, Common and Lesser Twayblades and Bird's-nest. It also has chapters on tropical species and (in the later edition) the Lady's Slipper. That was far more species than I would ever encounter in 'my' corner of England. If only I lived in Kent, one of England's orchid hotspots. Darwin is passionate, his experiments ingenious, his observations meticulous. George is the willing sidekick, but the orchids and the insects they recruit are the real stars.

By that point I knew that the Bee Orchid looks nothing like the Common Twayblade, which looks nothing like the Common Spotted, Lady's Slipper, Autumn Lady's-tresses or Lizard.

However, as Darwin makes clear, all these vastly different flowers, whether tiny temperate Bog Orchids with blossoms only millimetres across or tropical species like *Catasetum saccatum* with a trigger mechanism to fire pollen darts at insects, are variations on a common theme.

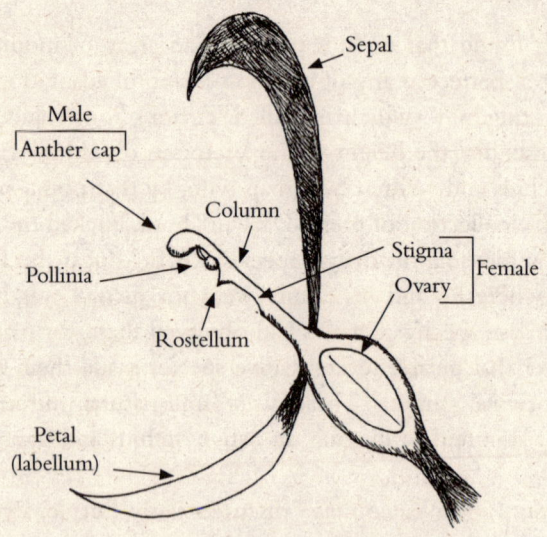

Cross-section of a typical orchid flower.

Whatever their shape, all orchid flowers are symmetrical. They have three petals and three sepals. In orchid flowers the sepals usually form an important part of the flower. As a general rule, as the flower forms it turns 180 degrees (a process known as *resupination*). This means orchid flowers are actually 'upside down', with a petal at the lower part. This petal usually becomes enlarged – this is the *labellum* (Latin for 'little lip'). Unlike the blooms of other plants, in orchids the external female and male parts are combined into a single organ, separated by a thin partition. This partition is the *rostellum*. The whole structure, containing rostellum, stigma, pistils and anthers, is the *column*.[17] The column, whether a tiny nub or a prominent beak, overhangs

the flower's centre. Often the column is tucked beneath a protective hood formed from sepals and petals. These are the caps of
England's 'little men' orchid species: Military, Burnt, Monkey,
Man, Lady. Finally, unlike other flowers, which produce dry
grains of pollen, orchid anthers usually carry pollen packets
(*pollinia*) bound together by elastic thread (*viscin* or *elastoviscin*).
Every one of these adaptations is crucial to successful fertilisation.
However, as singular as these characteristics are, more extraordinary secrets were yet to be revealed.

At that time, whether insects pollinated plants was still a topic
of debate. Darwin was convinced they did. One day, passing
some flowering Early Purples, he took out a well-sharpened
pencil and stuck it into the flower to imitate an insect probing
for nectar. When the tip entered the central part of the flower,
the anther (and the pollinia attached to it) separated from the
overhanging column. When he withdrew the pencil, it had two
little yellow blobs like tiny horns attached to it – 'firmly cemented
to the object [the pencil], projecting up'.[18] These blobs were
pollinia. Each pollinium was attached to one end of a little 'stick'
(Darwin called it the *caudicle*), which was fixed, via a gluey patch
at its other end (the *viscidium*), to the pencil.

This discovery provided Darwin with a eureka moment. He
glimpsed the complexity of the relationship between insects and
orchids in a way no one had before. Catching insects in the
vicinity of orchids validated his theory. Pollinia were stuck to
them, and not in a haphazard manner. A moth or bee flying
from orchid flower to orchid flower might have several pollinia
stuck to its proboscis, legs and eyes. Not only that, but they
were stuck in the optimum position for them to come into
contact with the corresponding orchid species' stigma. This was
a radically different mode of pollination from the more random
method of pollen transfer used by other flowers. It was also the
first of many more complex adaptations Darwin identified in
orchids.

Darwin observed that the tiny pollinia attached to his pencil moved. As the insect (or pencil) withdrew from the flower, the attached pollinia were standing upright. Soon afterwards, those vertical pollinia tilted forward. The process could take seconds or minutes depending on the orchid species and weather conditions. With the Early Purple it took under thirty seconds for the little horn-like pollinia to shift, not just slightly, but through a very obvious ninety-degree arc from upright to horizontal. Spellbound, Darwin investigated further.

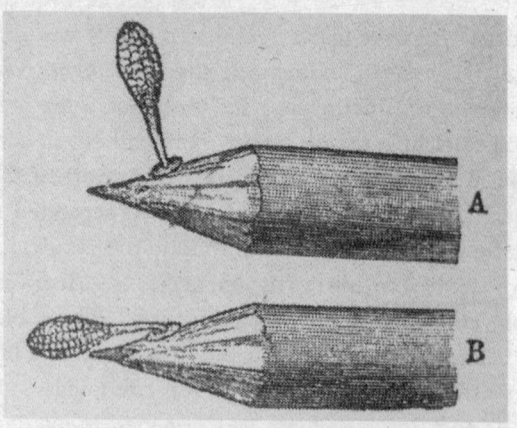

Dawin's pencil experiment.
Woodcut by Sowerby from Darwin's *On the Various Contrivances*.

He discovered that when it contacted air, the sticky viscidium at the end of the caudicle dried. As it hardened, the caudicle embedded in it (and consequently the pollinia stuck to the other end) changed angle from vertical to pointing forward. In general, the time it took for this change of direction to occur matched the foraging habits of the pollinating insect(s) used by that orchid. An insect could dip its proboscis into another flower – or several flowers – of the *same* plant in search of nectar, but during that time the pollinia would not yet be at the correct angle to contact the stigma of that species. The pollinia only reached the required angle after the average amount of time had passed for the insect to leave the pollinia-donating plant and start foraging for nectar on a *different* orchid plant. This was the proof Darwin needed to support two of his key ideas: first, co-evolution – insects and orchids must have co-evolved to develop this intricate mechanism – and, secondly, that this complex change must have developed to increase the chances of 'intercrossing', because intercrossing (Darwin believed) produced healthier offspring.

When he started to look at other species he found that each had its own peculiarities. On a cliff top above Torquay, Darwin observed that the small white-green trumpets of Autumn Lady's-tresses open in a spiral up the flower spike so that the longest-opened flowers are nearest the ground. Their pollinators, bumblebees, always alight at the lowest flowers, then crawl spiral-wise up the spike. When it first opens, an Autumn Lady's-tresses flower has a very narrow passage between the column and the labellum where the insect's proboscis enters seeking nectar. That confined space ensures that the proboscis is in exactly the required position for the pollinia in the overhanging column to stick to it, then get removed as the proboscis withdraws. The passageway remains narrow for a couple of days after the pollinia have been removed, too narrow for a proboscis with pollinia stuck to it to enter. This means newly opened flowers cannot be fertilised and makes it impossible for a bee which has just received the pollinia from that flower to reinsert them – not only into the same flower but into any flower on the same plant.

As the bee crawls up the flower spike, pollinia stuck to its proboscis, the passages of all the flowers that have only recently (or not yet) had their pollinia removed are too narrow for its proboscis to enter. Reaching the top of the flower spike, still seeking nectar, the bee flies off to try its chances at another honey-scented orchid nearby. There it follows the same foraging pattern: starting at the lowest flower in the spiral, it crawls upwards. If any of the lowest flowers on this plant have already had their pollinia removed, the narrow passage to the nectar will have, time allowing, widened enough for the bee carrying pollinia from a different plant to insert its proboscis. In the process, some of the pollen from the pollinia stuck to its proboscis will transfer to that flower's stigma. The bee removes its proboscis and makes its way up the flower spiral, inserting its proboscis, pollinia probably still stuck to it, into successive flowers. Each time it enters the passageway of a flower which has already transferred its pollinia, some pollen sticks to the stigma. At some point the pollinia will fall off. After that the bee will encounter another flower, higher up the flower spike, one that has recently opened, and there, in taking on nectar, that flower's pollinia will become stuck to the proboscis. So the process continues, a kind of relay race, with one runner – the bee – picking up and handing over pollinia from a new flower to older flowers on different plants.[19]

The dangling lime-green blooms of the Common Twayblade look a world away from those of Autumn Lady's-tresses: they are not bell-shaped, do not open in a spiral and do not have a narrow passageway for a proboscis to enter. Bumblebees don't visit them; small flies do. For this reason, their method of pollination needs to be different. Instead of having caudicles with sticky discs at one end and pollinia at the other (probably too tricky to attach to a tiny insect), the slightest contact (according to Darwin, 'a touch from the thinnest human hair') causes the rostellum of the Common Twayblade's flower to temporarily split.[20] In doing so, it curves forward and releases a drop of gluey fluid. The pollinia hanging from the column attach to the now glue-covered insect. The drop hardens, cementing the pollinia to its minute insect visitor. The forward curve of the rostellum shields the flower's stigma from these pollinia, preventing self-pollination. Over the subsequent few hours, the rostellum returns to its original position. By that time, the little insect forcibly conscripted into being the orchid's pollinia-carrying lackey will have moved on to a different Common Twayblade plant, and the pollinia will be transferred to a flower which has already lost its pollinia, one where the rostellum has straightened, allowing access to the stigma.[21]

The impressive Lady's Slipper flower gets its name from its enlarged labellum, which forms a pouch the shape of a shoe or slipper. Attracted by scent, bees enter the pouch, but they can't fly or crawl out by the same route. Darwin studied the efforts of a small pollinating bee:

> The bee vainly endeavoured to crawl out [of the flower] again the same way by which it had entered, but always fell backwards, owing to the margins [the inner margins of the labellum] being inflected. The labellum thus acts like one of those conical traps with the edges turned inwards, which are sold to catch beetles and cockroaches in London kitchens . . . Ultimately it forced its way out through one of the small orifices close to one of the anthers, and was found when caught to be smeared with the glutinous pollen.[22]

Several times Darwin dropped the hapless lab-assistant bee back inside the labellum and observed the same process: the bee could only escape by crawling through one of the openings in the back of the Lady's Slipper's pouch. On the way it got coated in pollen.

> I afterwards cut away the labellum, so as to examine the stigma, and found its whole surface covered with pollen. It should be noticed that an insect in making its escape must first brush past the stigma and afterwards one of the anthers, so that it cannot leave pollen on the stigma, until being already smeared with pollen from one flower it enters another; and thus there will be a good chance of cross-fertilisation between two distinct plants.[23]

Other British orchids offered their own curious contrivances, but whatever their complexities, it was clear they required a pollination technique far more refined than the use of a scuzzy paintbrush. Some of them even challenged Darwin's theory. Among these were the insect-shaped (*insectiform*) flowers of the Spider Orchids, Fly and Bee. Darwin had no doubt they were designed for pollination by insects, but some of those species, in particular the Fly, were not very efficient at attracting them – they formed very few seed capsules. If the small amount of seed produced was sufficient to replenish the species' population,

Darwin wondered why they produced so many flowers; on the other hand, if they produced the requisite number of flowers, why weren't more seed capsules formed? 'Something seems to be out of order in its mechanism,' he concluded.[24] Was its lack of fertilisation 'due to the proper insects having become rare under the incessant changes to which the world is subject; or to other plants which are more highly attractive to the proper insects having increased in number'?[25] In either event, the Fly did not appear to be honed to maximise reproductive success.

In an effort to solve this puzzle, Darwin turned to the Bee, but rather than answers, he found flowers which appeared to undermine his whole theory about cross-fertilisation. The Bee was 'excellently constructed for fertilising itself'.[26] Its pollinia detach from the column all on their own and dangle down to the flower's own stigma. Undeterred, Darwin insisted that, alongside a few other self-fertilising orchid species, the Bee Orchid was *originally* designed to be cross-fertilised by insects, but for some reason – perhaps absence of pollinators – they had evolved to fertilise themselves.[27]

Darwin's critics seized upon his use of the word 'contrivance' and wondered how there could be a contrivance without a grand contriver.[28] If a God had created all life, reasoned Darwin in return, why would that divine Being make so many mistakes and have to subsequently make adaptations to fix them? Surely, these complex plants are products of a 'long course' of natural modifications co-evolved with insects, which has led to divergence from a single ancestral model. Orchids, he argued, are the strongest living proof of natural selection.

Assisted by more sophisticated technology, researchers have since found flaws in some of Darwin's ideas about orchids. Darwin insisted that bees were too intelligent to fall for a cheap ruse like 'sham' nectar (which had been suggested by Sprengel)[29] and maintained that nectar was the only reason insects visited flowers, but it is now known that isn't the case. Many orchids shamelessly deceive pollinators with the promise of nectar which isn't there, and insectiform orchid flowers, like the Bee and Fly, take this deceit further.[30] Not only do they look like insects, but their flowers imitate the texture of insects and release cocktails of over a hundred different chemicals to precisely mimic the pheromones of a specific female insect.[31] This attracts male insects to them in an attempt to mate (*pseudocopulate*). During these amorous activities the flower's pollinia attach to the pseudocopulating male bee, fly or wasp and get carried to another pseudo-mate as the insect decides to try his luck elsewhere.

Despite errors like these, Darwin made some astoundingly accurate predictions. He was correct in supposing that the self-pollinating Bee Orchid originally evolved to be pollinated by insects. Where the long-horned bee (*Eucera longicornis*) still exists, the male swoops down to mate with the Bee flowers.[32] Similarly, while studying the mysterious, 30-centimetre-long nectar tube of the Madagascan Star Orchid (*Angraecum sesquipedale*), Darwin predicted a moth with an extremely long tongue *had* to be its pollinator. The moth was discovered twenty-one years after Darwin died.[33] Darwin knew nothing of DNA and genetics, but his suggestion that all orchids developed from a single ancestor seems to have been proven. Recent research into orchid genetics indicates the emergence of a proto-orchid about a million years before the extinction of the dinosaurs. My nine-year-old self, standing in awe of that first tropical orchid I saw, felt vindicated: it *had* been teleported (albeit over 67 million years and countless gradual modifications) from a steamy Cretaceous jungle.[34]

All orchids derive their identifiable characteristics from that ancestral orchid. Flower symmetry, their (usually) 'upside-down' flowers, the 'upper' petal forming the labellum (landing platform, insect trap, trigger mechanism or a decoy insect); the pollinia, packaged and protected in a way which reduces waste and which can pollinate multiple flowers; their dust-like, easily distributed seed – all of these characteristics helped those ancestral orchids survive the mass extinction of 66 million years ago, which marked the end of the age of dinosaurs and eliminated 76 per cent of all other life on Earth.

Winter. Of my guerrilla orchids in the city, nothing showed. In the corner of Dad's garden a couple of leaf rosettes splayed across mossy earth. The Bees had been stirring since late autumn. I kept a hopeful eye on them, awaiting flower spikes which, I now knew, would pollinate themselves. For months, they did nothing. The leaves frayed around the edges from frost and grazing slugs, waiting out the winter with more patience than I.

As the days warmed, lilac tufts of bugle spread across the meadow. A swaying green haze where sedges, fescues, dog's-tails and meadow grasses hoisted their slim flowers followed. White clover opened bushy orbs, buttercups lifted their glossy dishes to the sun. Flies, bees and beetles crawled through their pollen and common blues arrived to wobble through the air on sapphire wings. The Bees started growing again. Fresh lime-green leaves pushed upwards and from the sheath of a central upright leaf (the *bract*) a parcel of heavy buds emerged. I counted six plants. Only two had buds.

One evening in June, the lowest flower opened. It had been promising to for days. The bud had twisted to the horizontal and flushed from pale green to white, to pale pink, but its sudden opening was still a surprise. Like a newly minted big brown-bottomed velveteen bee decorated by golden curlicues it appeared, sitting between three flakes of pale rose. A few days later, the next flower up the stem twisted, dipped, faded, opened. Then the next, and the next, and the next, each haloed by pink pennants, each overhung with a slim green beak which soon dangled a pair of sun-yellow pollinia.

Those two plants produced twelve flowers. The dozen little jesters grinned at me, but now it felt as if I was in on their joke. Those cheeky flowers were upside down. The 'beak' was the flower's column. Two of their petals had become slim antennae-like structures, the third formed the hairy bee-like 'body' and it released waves of kairomones, formed from over a hundred different chemicals, calling to a species of solitary male bee which had died out locally thanks to humans. In that bee's absence, those flowers had 'learned' to devour part of their own column so that those baubles of pollinia could droop far enough to reach the flower's own stigma.[35]

June became July. Without a single insertion of a pencil tip, the Bee sepals lost their lustre and collapsed inwards; the velveteen labellum withered, the pleated ovaries swelled.

In the playground, a churchyard and a park a scattering of guerrilla-planted Common Spotteds splashed the brick-and-concrete world with pink, white and lilac botanical complexity.

I wondered about the practicalities of using those plants as orchid orchards. I didn't fancy crouching next to them as I tried to coax tiny pollinia from one flower onto my pencil/cocktail stick/needle, then rub them against a tiny stigma in full view of commuters, kids, families and the homeless. Besides, even if I managed to pollinate them, they might get picked, crushed or poisoned. Orchid orchards, I conceded, were best kept protected and private.

Fortunately, I probably already had what I needed, rescued earlier in the year from a bramble-covered patch in a corner of pastureland destined to be concreted over. Even before diggers moved in, starved of light and pollinators, the colony of Common Spotted I had found were already showing signs of being the collateral damage of an abandoned countryside. Scrub had encroached. Without intervention, the orchids were struggling to bloom. In a season or two, their new tubers, unable to receive enough nutrients, would have become smaller and smaller, the plants weaker, until they failed to bloom and, ultimately, died. I rescued six plants, all with flowers (even if rather small flower spikes), and potted them up in my yard. I would use them as my Common Spotted Orchid orchard. At least, that was the plan.

With Darwin's orchid book as my manual, I prepared my tools – pencils sharpened until their graphite tips could pass for black fish bones; a needle, a few bristles of a paintbrush, a cocktail stick, anything that could pass for a proboscis – then placed two of the flowering Common Spotteds in their pots on a table in front of me and sat down eye to eye with them, preparing to do what Darwin had. Sort of.

Darwin wrote about many orchids but not the Common Spotted. I considered the small pink-and-white speckled flowers in their feather-duster-like formation, their darker black dashes like an indecipherable code. I'd use his observations about Early Purple, Pyramidal and Fragrant Orchids (they all had a similar flower structure to the Common Spotted) and see what happened. I picked up the needle and peered into the shadowy passage at the centre of one flower. I needed to trigger the release of the caudicles and allow the viscid discs to cement onto the needle's shaft, where they would (I guessed) arc forward ready to be transferred to flowers on another plant.

I wondered what advice Darwin would have offered. He probably would have said, 'Just have a go.' So I did. I had dozens of flowers to practise on. The needle entered, then, as if withdrawing a diminutive Excalibur, I brought its steel shaft back into the light. Nothing stuck to it. I tried again. Nothing. Again. The same result. Wondering whether an insect had already taken that flower's pollinia, I raised the flower to peer at its column to see if I could find any. I should have started with that. My guides to what I was looking for were a few 150-year-old engravings, yet when I looked into the flower I found matching anatomy. There was the rostellum. There, slipped into their own special cavity on the column, a pair of pollinia, snug as peas, shaped like green commas.

Darwin had discovered there was always a trigger to release the pollinia. Sometimes it was on entry, sometimes on the insect's withdrawal. Nothing happened going in, but when I moved the needle up and back, a tiny antler sprang out of its cavity onto the needle. As Darwin must have done, I watched, mesmerised, as that tiny horn and its little green pollinia slowly moved, as if by magic, through an arc of several degrees.[36] It was graceful, precise, incredible, just as Darwin wrote: 'A poet might imagine that whilst the pollinia were borne through the air from flower to flower, adhering to an insect's body, they voluntarily and eagerly placed themselves in that exact position, in which alone they could hope to gain their wish and perpetuate their race.'[37]

Exhilarated, I moved the needle, pollinia attached, carefully through the air until I could insert it point-first into a flower on the other plant. So that I too could 'perpetuate their race', I lined up the pollinia with the flower's stigma, did my best imitation of a foraging bee and rubbed the green blob across the surfaces of half a dozen stigmas before the pollinia and caudicle on the needle fell off.

Satisfied, I sat back. What would Darwin have thought? It seemed to me that if he knew that all of the Pyramidals, Bees and Autumn Lady's-tresses he and George had spent that summer studying had died out around Torquay many years ago he wouldn't much mind me using his knowledge to try to stem the rapid disappearance of the plants that were his passion.

9

BURIED HORIZONS

My first orchid seed came from the Bees in the mini-meadow. Over six weeks I watched each ovary swell into a pod about three centimetres long, thick-skinned and leathery, like large, green cardamom pods. Slowly they changed from green to yellow to patches of brown, until they started to look dry and brittle. By then, much to my parents' curiosity, I was visiting the meadow daily. On one occasion in July I noticed that a pod had a little split along a lengthwise pleat. I prodded it. A puff of brown smoke. The smoke dissipated. Seed. My heart yo-yoed. There was dust-like seed! (Yay!) It was blowing away. (No!) I plucked the pod from the stem (more smoke), cupped it in one palm and teased it open.

Opening a cardamom pod reveals a dozen or so small, stony, black seeds neatly arrayed in rows. Inside the Bee pod a layer of specks coated the inner lining, but it was otherwise empty. Scraping a fingernail across those specks pulled some of them away. They danced into the air; a few landed on my skin. I fished a tissue out of my pocket, laid the pod inside and attempted to brush the flecks off my palm onto the tissue. They stuck to my skin, got under my nail, drifted away on the breeze. When a gust tumbled the tissue and empty husk away, I aborted the mission. There were other plump pods on that stem. A couple had not split. Carefully, I separated them from the stem and left the others (maybe I would need more in the future; maybe I was picking them too soon). Cautiously elated, I carried them home. 'My' Bee Orchid orchard had provided a harvest that could lead to dozens of new plants. The thing was, getting new plants involved steps that were not going to be easy; they were also steps that flipped my understanding of the world.

To anyone peering through the kitchen window that warm summer evening I probably looked like a drug dealer as I stooped over the kitchen table above a sheet of A4 paper folded once down its length then laid open again. I was now wearing a face mask to reduce the risk of breathing on and scattering that fine seed. My plan was to open the ripe pods, extract the seed onto the paper, then funnel it carefully along the prefolded crease into a waiting test tube for drying.

The pods had a heavy musk-meets-clove scent. With a thumb-nail, I split the first one. Brown dust foamed out. I quickly learned that the patience of a saint, fingers as numerous and thin as spiders' legs, the steadiness of a sniper and nostrils on the back of the head would be a useful combination when dealing with orchid seed. For a big-handed oaf like me, getting that seed where I wanted – onto the paper, then into a test tube – was like herding a million reluctant, near-microscopic cats.

'Million' and 'microscopic' are not exaggerations. Of all flower-ing plants, orchid seed is among the smallest and produced in the largest numbers. Any orchid's seed capsule contains a clump of brownish dust (red-brown, green-brown, yellow-brown, dark or pale brown; sometimes more elongated or more spherical, depending on species). It makes a grain of rice look gigantic. Any ill-timed sigh will spirit it away, dispersing it like minute balloons carrying embryos into the world. And there are thousands of them. Darwin estimated that a Heath Spotted Orchid produces 186,300 seeds every year.[1] I wonder if he ever actually counted.

When I tried to get the Bee seed where I wanted, I failed. It didn't want to be coaxed via folded paper into glass test tubes.

It tumbled everywhere. In those first forays I set out to save every dot. I was destined to fail in my objective – besides, it was unnecessary: a quarter of a teaspoon of seed, about a fifth of the seeds in a single Bee capsule, is more than enough to germinate dozens of plants – if everything goes according to plan. Often it doesn't.

Darwin indicated just how frequently things don't go to plan when, allowing for one acre of land to hold 174,240 plants, he determined that within five years the descendants of a single Heath Spotted Orchid *should* cover 94 per cent of the world's land mass, but . . . they didn't. Many orchids were already rare in Darwin's time and he was not sure why so few seeds germinated. 'What checks the unlimited multiplication of the Orchideae throughout the world', he admitted, 'is not known.'[2]

Other nineteenth-century orchidologists grappled with the same question. Some noticed that tropical seeds occasionally germinated in pots around the parent plant, but trying to grow orchids elsewhere, let alone on a potentially lucrative scale, appeared to be impossible.[3] This posed a conundrum: orchids produced huge amounts of seed but it hardly ever germinated, yet there *were* orchids in the world, so they germinated somehow – but how?

Orchid seed itself offers a clue. My microscope cost £8 on eBay. It looks as if it was salvaged from a school in the 1950s, but it's all I need to see the secrets of seed. When I first looked at a few Bee seeds through the eyepiece, I found rice-grain-shaped asteroids, their surfaces pocked and cratered. They looked as if they had been sculpted out of brown glass. The seed's outer coat (*testa*) was so thin that I could see the orchid embryo through it, like a dark orb. When I angled the mirror to reflect light from directly behind the seed, the testa became almost completely transparent, the thicker sections silhouetted like a gothic cage around the embryo, a black jewel.

Those craters in the testa form air pockets capable of carrying the extremely light seed for miles on a breeze. They can also help it float long distances on water, and being an uneven surface, it can cling to fur, feathers, hair and mud on hooves or footwear. This explains why some British golf courses are among the best places to find orchids: first, the land is left largely undisturbed and undeveloped, with scrub not allowed to encroach; secondly, seed is carried there from other golf courses on the soles of shoes or the tyres of golf trolleys.

In terms of understanding why the world is not overrun with orchids, what matters isn't the testa so much as what's inside, or rather, what isn't. An orchid seed consists of the testa, a simple embryo and . . . nothing. The seeds of sunflowers, geraniums, bluebells, cress – pretty much any other flowering plant or tree – contain not only an embryo but also a fertilised endosperm. The endosperm provides enough nutrients to the embryo for it to emerge from the ground and begin photosynthesising. This is why, with most plants, if you put a seed in moist soil, the seed begins to swell, the testa softens, the food store inside the seed is metabolised, the seed coat splits and the first root (the *radicle*) emerges. At that point the growing embryo starts to become self-sufficient. The endosperm has done its job. Not so with orchids.

To stay so tiny, orchid seeds have basically no endosperm. They are cast out from the seed pod, essentially prisoners that, once awoken by the right conditions, must escape their testa prisons without being provided with enough prepacked supplies to do so. That's the check on rampant orchid spread which Darwin, his focus on flowers rather than seeds, didn't know about. As a 'check' it's also a riddle. How does the seed break out? Where does it get its energy from? It took a chance encounter in 1899 to find the first clues to solving this puzzle.

In 1899, Noël Bernard, a French botanist with smouldering looks and a neatly waxed moustache, was serving in the army near Fontainebleau Forest. On 3 May he went for a stroll. There, in the forest, he happened upon some orchid seedlings growing where an orchid flower had become covered in leaf litter.[4] The flower and seedlings were of the Bird's-nest (*Neottia nidus-avis*), an orchid also native to Britain.

Early herbalists were suspicious of the Bird's-nest. Gerard remarked that it is a 'bastard' and 'degenerate kinde of Orchis, and therefore not used' in medicine.[5] Centuries later, even Darwin called it an 'unnatural sickly-looking plant'.[6] To be fair, even amid the vast physical diversity of the orchid family, the Bird's-nest is an outlier. Only if conditions are right does it appear, usually around late May after a damp spring, when it pushes frail, old-bone-coloured flower spikes up through fallen litter in gloomy woods where few other plants grow. It appears to be leafless; it isn't green; it has no chlorophyll, so can't convert the sun's rays into energy.

This combination – a plant growing in low light without the ability to turn sunlight to energy – led Bernard to wonder how those young orchids survived. He performed a series of experiments and discovered that the germinating Bird's-nest had been colonised by a fungus. Oddly, though, the fungus wasn't eating the orchid; the orchid was consuming the fungus. That was how it survived without sunlight, but how was it possible?

By 1899 it was known that many plants and trees grew alongside specific soil fungi called *mycorrhizae* (singular *mycorrhiza*).[7] Mycorrhizae form a vast web of fine filaments (*hyphae*, singular *hypha*) through the earth. Plugging into this network allows plant roots to absorb more nutrients and boost their resistance to disease and drought. It's highly likely that the mycorrhizae benefit too. An exchange of nutrients appears to be key: plant roots recruit a fungus and bind to it (or indeed the other way round); the fungus gains nutrients from the plant it cannot otherwise obtain (including vitamins, amino acids, carbohydrates), and the fungus provides important minerals in exchange. Although the precise details remain unclear, the benefits are mutual (*symbiotic*). But where does an orchid devouring a fungus instead of sunshine fit in this plant–fungus world?

Although it may sound odd, the Bird's-nest is not alone in snacking on fungus. Since Bernard's discovery, 170 achlorophyllous (having no chlorophyll) orchid species have been found, all of them dependent on fungus. This is the largest number of such species in any plant family.[8] Three of them live in Britain – the Bird's-nest; the Coralroot (*Corallorhiza trifida*); and the Ghost Orchid (*Epipogium aphyllum* – *aphyllum* from the Greek for 'leafless', because it appears to have no leaves; *epipogium* from the Greek for 'overbeard', because the flower's labellum, uncommonly for an orchid, overhangs the rest of the flower, which dangles down in a way someone must have thought resembled a ragged beard). To me the flowers of the Ghost look like small, pale, pink-white squid. Like most people, I can only make this comparison based on photos because, if it's not already extinct, the Ghost is Britain's rarest plant.

Darwin makes a single reference to the Ghost and notes that it has appeared only once in Great Britain.[9] That was in late July 1854, when Mrs Anderton Smith spotted an unusual-looking flower growing in a muddy Herefordshire lane. She picked it and sent it to a botanist to learn what it was. That botanist (the Master of Taunton Grammar School) posted it on to a friend of Darwin, Hewett Watson, a brilliant (if irascible) scientist. Watson identified the specimen as *Epipogium aphyllum*.

A few weeks after his wife made the discovery, Reverend Smith went searching for more of the mysterious flowers. He found a 'considerable' number growing where woodland had been cleared and the earth churned by the hooves of horses dragging the timber away. Not having to worry about laws designed to protect rare species, the Reverend dug up a few of the plants to embellish his garden. They didn't survive there long. At the time, no one knew that the Ghost, like the Bird's-nest and Coralroot, is a thief and that snatching the thief away from its victim will kill it.[10]

The intricacies of this thief-like relationship are yet to be fully unravelled, but it seems that somehow, via a specific mycorrhiza, these orchids steal nutrients from nearby trees (usually beech, willow and oak).[11] While the tree and the fungus, and the orchid and the fungus, have symbiotic relationships, in stealing nutrients from the tree via hyphae, the sneaky Ghost, Bird's-nest and Coralroot are parasitic on the tree. It's a *tripartite* relationship, not a simple orchid–fungus nutrient exchange; a third party (the tree) is involved.[12] So it was that, wrenched from the mycorrhizal network, the Reverend's Ghosts did not survive. Others later searched for the Ghost in the same location, but 'Eppie' had vanished.

This pattern – discovery, often by lone women, identification, then destruction by collectors for gardens, herbariums and as museum curiosities – was repeated on the few occasions when the Ghost materialised in the Chilterns and Welsh Borders. Invariably it then vanished, sometimes for decades. After 1854, confirmed sightings occurred in 1876, 1878, 1892, 1910, 1923, 1931 and 1933. Another long hiatus followed before it reappeared in 1953. Until 1987, its rarity seemed to decrease, with a few dozen flowers occasionally appearing. Then it disappeared. Again. With no corroborated records for twenty-three years, in 2009 the charity Plantlife declared the Ghost extinct in Britain (the IUCN Red Data List continues to class it as Critically Endangered) and used its fate to raise awareness of Britain's beleaguered wildlife. With audacious timing, unknown to Plantlife, days before the declaration, a determined amateur botanist had found the Ghost in the Welsh Borders. At the time, very few people were told.

Despite committed searches by teams of volunteers, to date 2009 marks Eppie's last appearance in Britain. Given that decades often pass between sightings, there's still a chance its wraithlike, squid-shaped banana-smelling blooms might reappear.

While Noël Bernard's initial research showed that the Bird's-nest is *entirely* dependent on mycorrhizae, his subsequent investigations found that all native European orchids have a similar tendency.

They may not take it to the extremes of the Bird's-nest, Coralroot or Ghost, but they all need specific soil fungi – orchid mycorrhizal fungi, different from the mycorrhizae used by other plants – to survive. Without the right fungus, the embryo cannot develop. If the right mycorrhiza is present in the earth when the tiny seed enters the soil, there, in darkness, hyphae penetrate the testa through incredibly small holes and enter parts of the orchid embryo. Inside the embryo the hyphae coil up into clumps of closely packed whorls called *pelotons*. The embryonic orchid digests these to gain energy and micronutrients.

Although the pelotons are consumed to power the growth of the embryo, it appears the fungus gains something, perhaps carbohydrates, from the orchid, so the fungi that orchids recruit are known as *symbionts*. Some researchers go further and use the term *mycobiont* to describe the fungus.[13] This name is normally used for lichens. Lichens are composite organisms, comprised of algae or cyanobacteria *and* fungi growing together, completely interdependent. The degree to which the fungus in the orchid–fungus relationship is dependent on the orchid is not clear, but together they too create a kind of composite organism, fungus + orchid, and this mutually beneficial union is what allows the orchid to grow into the next phase of its development: a small structure found in no other family of flowering plants, a *protocorm*.

A protocorm is a lump of cells a few millimetres long, essentially a vehicle for fungal infection. A small, slightly lumpy hard blob, it grows fine hairs into the surrounding earth, allowing more of the symbiont to grow into it. It also excretes fungicidal substances to control the fungal activity occurring within and around it. Under normal circumstances, many orchid fungal partners are parasitic or pathogenic; the protocorm's excretions subdue this behaviour and allow the mycorrhiza to harmlessly grow into it.[14]

Living off these hyphae for months or years, the protocorm grows. Over time its cells form the roots, tuber (or rhizome) and leaves of the orchid seedling. As it matures, the seedling may switch its preferred species of fungus dinner (a process known as *fungal succession*), because the symbiont used in early stages of orchid development is not always the same required by older orchid plants. Exactly how orchids make this switch and why is another unknown, but throughout their lives European orchids maintain a dependent relationship with mycorrhizae. Orchid species that fully die back and lose their roots and leaves from season to season recruit new fungal partners from the soil each year as they break dormancy; for some species, the mycorrhiza they need lives on and in the skin of the tuber 'waiting' for the orchid to grow new roots, which the fungus can then recolonise.[15]

Given that Britain's orchids inhabit infertile soil, these fungal partners play a crucial role in providing them with the minerals the orchids would otherwise struggle to obtain. With the Ghost, Bird's-nest, Coralroot and several species of helleborine, this partnership enables orchids to colonise gloomy woodland where there is little need to compete with faster-growing green plants. On the other hand, if an orchid seed lands somewhere which appears to be ideal but the required fungus is not present, the embryo will die in its testa prison.

Discovering the secret of how orchid seed germinates – or, more often, doesn't – made it clear to me how a Common Twayblade manages to survive four years underground before appearing as a young plant; how the rare Monkey Orchid, with its candyfloss-coloured monkey-shaped blooms, manages to spend five years growing underground; and how other species – Autumn Lady's-tresses, Burnt and Lady's Slippers – can spend eight years or more developing as protocorms in darkness.[16]

It was also evident that orchids had led me to a place I hadn't really thought about before. Until learning about orchids' fungal partners, I had seen earth as a dead substrate. Eroded rocks, decayed things, handy for walking on, sticking plants in and for building the foundations for roads and buildings. Thanks to orchids, I started to see earth for what it really is: a three-dimensional living realm. Earth – soil, dirt – is *really* a living element, like a lightless sea or sky, coral reef or savannah made up of different habitats called 'soil horizons'.[17] These horizons are filled with vast shoals, herds, flocks of ancient microscopic life;[18] they are buried horizons where the Ghost has been said to flower and where mycorrhizae reach through layers of soil creating vast webs, feeding protocorms, and connecting orchids to trees.[19]

Learning this made me aware that Darwin's observations on pollinators and orchid flowers did not explore the whole picture. Without underground fungi there would be no complex blooms, no co-evolution with pollinators. Darwin was continuing a pattern of looking at the world that stretched back to early herbalists and beyond: *they* had focused on roots, rather than flowers or pollinators, and in that sense their understanding of an orchid was incomplete too. Gerard, Parkinson, Ray, Johnson, Smith, Curtis, Darwin and the rest, together with their often exquisitely illustrated studies, showed orchids in magisterial isolation, but none of them was right about what an orchid is, for an orchid in isolation is not all an orchid is.

Orchids co-evolved with insects to recruit pollinators, but of equal importance is their co-evolution with underground fungi. Orchids disproved the simple picture of segregation I had been taught: food chains, higher and lower organisms, complex and simple life: *they* were the lessons I had obediently learned for school exams, and they invariably placed humans apart from and superior to the rest of the living world. Darwin's theories slot into and perpetuate this view, but Darwin didn't know that orchids were dependent on fungi, that complex interactions took place in the earth around and within orchids' roots, or that they formed protocorms that were symbiotic fungus–orchid organisms.

I found myself wondering about the coils of hyphae within an orchid protocorm and about the orchid + fungus being they form, then I wondered about the human + bacteria + fungi + multiple other microorganisms being that is a person. I wondered whether many scientists had it wrong when they overstated the unusual symbiotic nature of lichen or mycorrhizae and orchids, as if they were curious exceptions. Really, it seemed to me, every being can be called symbiotic.

If Darwin had been aware of the complex relationships between orchids and the unseen ancient life of earth, indeed between all visible and invisible lives, might he have written his theory of natural selection, of the 'struggle for existence' and the 'extinction of less-improved forms' differently?[20] How might the world of the late Victorians (and ours too, for we have inherited their habits and ideas) have been different if, instead of humans at the pinnacle of an evolutionary pyramid, there was greater recognition that the pyramid, and each of us as a living organism, is held together by multitudes of other species, intertwined, within us, on us, around us, holding all life in a careful state of intertwined, symbiotic equilibrium?

GROWING DUST

The kitchen looks as if something is about to be hacked into bloody pieces. Plastic sheeting covers the floor and table. Everything has been wiped down with bleach as if to conceal evidence from crime-scene investigators. A pressure cooker hisses on the worktop. The outflow of an air-purifier is Blu-Tacked to a hole cut out of the bottom of a large plastic box lying on its side on the plastic-sheet-topped table. The purifier rattles as it exhales a steady stream of filtered air. Sitting in front of it is a bit like being face-to-face with a diminutive wind tunnel. This is what my research into orchid-seed germination has led to: me wearing surgical gloves, an apron and a face mask, holding a large chef's knife as I check that everything in this suspicious scene is ready.

As dubious as it looks, the only thing about to be hacked up is a chunk of swede. But where, alongside the plastic box, petri dishes and other bits and bobs, is the all-important fungus that orchid seeds need to germinate? I'm attempting to germinate seeds *without* their fungal friend. But after all I have explored and explained, how is that possible?

Noël Bernard's discovery that orchid seed relies on fungus transformed attempts at germinating it, but issues remained. Specific mycorrhizae had to be found: the right mycorrhiza to match the species of orchid. (Some orchid species, it's true, are less picky – they will create a union with a range of mycorrhizae; other species work with only a single species.) Finding the right mycorrhiza meant isolating the fungus from the roots of a living orchid, then growing the fungus in cultured form before combining it with orchid seed.

Isolating the fungus takes place in petri dishes containing agar mixed with oats. A little piece of orchid root cut from a living plant is dropped onto this, the dish is put aside and, over time, the fungus (or fungi) growing in and on that root spread through the surrounding medium. The technical term for this fungus-isolating technique is 'fungus baiting'. Once the hyphae are in the medium, that bit of medium can be removed and added to other petri dishes. The fungi will then colonise the medium in those dishes. When seed is added, with luck, the fungus combines with the embryos inside the seeds.

Baiting fungus and cultivating it isn't very difficult. People have been nurturing fungi for millennia to make bread, inoculate blue cheeses and brew beer. In London in 1909, Joseph Charlesworth, a tropical orchid importer, teamed up with a mycologist, John Ramsbottom, with the intention of isolating the right fungus to germinate orchids and then embark on lucrative mass orchid production. Ramsbottom isolated a fungus and for the first time in history tropical orchid seed could be germinated on a large scale. However, their success, although considerable, was incomplete. Only some orchid species responded to the mycorrhiza being used. Ramsbottom and others were unable to establish which fungus the seeds of non-responsive species needed; the fungi isolated from their parent plants and from other species didn't lead to germination.[1] Without the right fungus, living plants of those species could still only be obtained from the wild.

About a decade later, Lewis Knudson, an American professor of botany, explored the idea of bypassing the orchid seed's requirement for a fungus.[2] Knudson hypothesised that the seed doesn't really need a fungus so much as the substance the embryo gains from the fungus. He believed that substance could be a carbohydrate, so he compared the germination rates of tropical orchid seeds planted on peat with those planted on sugar cane. The seeds on peat showed no sign of germination; those on sugar cane did. Knudson experimented with orchid seed and different amounts of sugar cane extract. The results led him to create a kind of holy grail for orchid growers, a mixture he called Knudson B. Combined with an agar solution in sterile conditions, Knudson B became a nutritious jelly. Tropical orchid seeds planted on it would germinate, become protocorms, then seedlings; those seedlings could be removed from the sowing medium and planted in bark and/or moss. Hey presto, orchids *without* the hassle of isolating and propagating fungus!

Knudson had invented *asymbiotic* germination (germination of orchid seed *with* a fungus is known as *symbiotic* germination). The asymbiotic method allows orchid embryos and protocorms to absorb the nutrients they need directly from the growing medium. This technique would never work in the wild because levels of micronutrients are small and sufficient concentrations are only available via a fungus. Ultimately, Knudson's B formula worked well with some orchid species but not others. He altered the ingredients and settled on a solution with added copper, manganese, iron and zinc. This – Knudson C – proved to be even more potent, offering successful germination across a wide range of tropical orchids.

Since Knudson's discovery that tropical orchids could be asymbiotically germinated, research into the role of mineral nutrition in germinating tropical orchid seed has led to many different growing solutions, such as P723, Murashige and Skoog, Fast's, Norstog, Hill's, Mead and Bulard, Vacin and Went. Each of these has varying concentrations of micronutrients and minerals, for it has been shown that a very small change in ingredients can have a significantly beneficial impact on the germination of some orchid species but negatively impact the germination rates of others. The implications of this for growing tropical orchids for the houseplant market and conservation purposes are significant, and many studies have been published comparing the success rates of different orchid growth media on different species.[3] For much of the twentieth century, most of those experiments and most conservation interest focused on tropical species. It took a while for researchers to shift their attention to Europe's terrestrial orchids, which, while everyone's scrutiny was on the plight of rainforests, had been rapidly vanishing. Suddenly the question arose: how did *their* seeds respond to Knudson C?

I don't know Svante Malmgren. At the time of writing, this guru of growing native (particularly northern) European orchids is alive and well, in Sweden. An online photo shows him to be a mild-mannered optician-type, bespectacled, clean-shaven, grey-haired, unaffiliated to any university or botanical garden. While others have focused on tropical species, Svante appears to be a guy with a passion for Europe's orchids.

He has spent over twenty years successfully growing them in considerable numbers in his garden or under lights in his cellar, sheltered from long Scandinavian winters. Intent on saving dwindling numbers of native Swedish species, Malmgren discovered that Knudson C had limited effectiveness in germinating them, so he created his own, more successful, growing medium for terrestrial orchids: Malmgren's Medium.

Dedicated to orchid growing, Malmgren's website describes techniques to germinate various European species, including the Bee Orchids of which I now had several thousand seeds. Following his advice, I set out to germinate them. For the medium, all I required was a packet of powdered Malmgren's Medium, distilled water, agar and (weirdly) a few bits of pineapple juice, potato or swede. In the way of equipment, I needed some petri dishes, a measuring jug and precision scales, sterile syringes, test tubes, a microwave, a pressure cooker (for sterilising things – laboratories use something called an 'autoclave', which is essentially a big pressure cooker that costs ten times as much), a blender for mixing the ingredients, surgical gloves, a face mask and an air-purifier hooked up to a plastic box. And a fridge.

I got most of what I required off the Internet, much of it second hand. One packet of Malmgren's made one litre of medium, enough for twenty or thirty petri dishes. That number could lead to around two or three hundred orchid protocorms. The total cost came to significantly less than the price of a return train ticket to London. I felt it was a reasonable investment. However, getting the equipment and ingredients was the easy part.

The complexity of what I was facing struck me that day in October when I set everything up in the kitchen and surveyed my home-made mini-wind tunnel, breathed the bleach fumes and mentally checked off everything I needed. It felt a bit like embarking on a game like chess, without knowing all the rules. Only one thing was clear: I had multiple opponents. They were in the air, on every unsterilised surface, in the tap water, on my breath, on every skin cell and hair I shed. The symbiotic relationship between orchids and mycorrhizae had opened my eyes to a world of unseen underground life; attempting to germinate orchid seed was making me equally aware of just how much unseen life surrounds us.

Trying to germinate orchid seed in an asymbiotic way meant keeping everything sterile. *Everything.* If not, contamination would follow. Contamination meant death to orchid embryos. Trying to ensure my kitchen remained sterile for the duration of that seed-sowing episode was going to be a challenge. It made me appreciate just how misguided the view is that people are somehow masters of this planet. Really, we lurch through it, clueless about what we tread on, prone to ignoring that each of us is a galaxy of microscopic beings, microhabitats, even as we breathe microbes in, exhale them, are smothered in them, depend on them. They and their spores float in the air, fall down in a slow rain to be kicked up again by every movement we make, and if any happen to settle on a petri dish full of Malmgren's Medium they will colonise and feed off it, multiply, outcompete and smother the orchid embryos within.

Laboratories counter microbes by creating sterile work areas using glove boxes (those glass tanks with a couple of rubber gloves extending into them where everything inside the box is sterile) or large contraptions called laminar flow cabinets which blow air purified through very fine filters to remove dust, fungi, bacteria and viruses over a workspace. A glove box or lab-quality laminar flow cabinet would be great. Unfortunately, they cost thousands of pounds each (second hand), which is why I had a domestic air-purifier Blu-Tacked to a hole in the bottom of a 60-litre plastic storage box: not pretty, not robust, not laboratory-grade, but hopefully capable of filtering out enough microbes to keep the box reasonably contaminant-free.

Anything entering the box needed to be sterilised. That was where the pressure cooker came in. Everything not already sterile needed a visit to the cooker. I wrapped all the bits and pieces I thought I might need – glass beakers, metal spatulas, petri dishes, a bottle of distilled water for rinsing seed – in foil and crammed them into the little cooker, then secured the lid, put it on the 'soup' setting and waited. After twenty minutes of gurgling and steaming, it was time to let the contents cool before moving them in their foil packets to the plastic box (the foil was to protect them as I moved them through the microbe-packed air from sterile cooker to sterile box). Once in the box, I stood them in the steady breath of the purifier. It was like moving pieces on a chessboard; I had to plan each move in advance. I've never been good at chess.

The opening gambit was easy: in a big plastic measuring jug I made a soup out of powdered agar, distilled water, Malmgren's Medium (a very fine black powder), a little bit of pineapple juice and small chunks of swede. The ingredients had to be combined in the correct quantities – that's where those pocket-sized precision digital scales (which I assume are more often used by unlicensed dealers in controlled substances than by orchid micropropagators) come in handy for measuring 7 grams of powdered agar. As for the rest, a regular plastic measuring jug is fine for calculating a litre of distilled water and 20 to 40 millilitres of pineapple juice; a tape measure works for measuring 10 cubic centimetres of swede.

Add the Malmgren's and agar, blend it all together and you'll get a black soup which gives off hints of rank mushrooms and suspect herbal tonic. I used a digital pH meter (from a pet shop, designed to check fish-tank water) to test the pH of the soup. If it was much above 5.8, I had to add a drop of sulphuric acid; if it was much below, it needed a similar amount of potassium chloride. Check the pH again – aiming for 5.8, but anywhere between 5.5 and 5.9 was fine. Once the pH was right, I blended again, then put what was perhaps the world's least appetising soup into the microwave for a few minutes on high to break down the swede. As it warmed, the mushroom scent grew. *Ping.*

When cooked, pour the liquid from the jug into the petri dishes and put the dishes into the pressure cooker to kill any contaminants. It's possible to pause the game at this point, to take the filled petri dishes out of the cooker, let them cool and allow the mixture to solidify (the medium in the dishes needs to be cool and solid before you can add seeds). You can wrap the dishes in foil or clingfilm and set them aside on a kitchen shelf for a week to see if any contamination grows. If it does, discard the contaminated dishes. You then set up the whole chessboard again, but this time just to sow orchid seeds. Or you can plan to do what I did that day, which was to save the time and effort of dismantling everything and setting it up again by moving everything from the cooker to the microbe-shielded box.

Collected from brown, mature seed capsules (like my Bee seeds), orchid seeds should be dried for a few days. For most species the embryos inside keep their viability longer that way.

I dried the seed by funnelling it into test tubes and standing the tubes (without the tops on) in a bigger Kilner jar half filled with oven-dried rice. The dried rice absorbed the moisture from the seeds. Then I removed the test tubes, replaced the airtight tops and stored them in the fridge. Just before sowing, the seeds from those test tubes need to be sterilised, partly because seeds taken from a dried capsule will be coated in microbes, so putting them into the growing medium without sterilising them will result in contamination; partly because sterilising the seed helps break down the tough seed coat. This allows nutrients from the growing medium to seep in and feed the embryo.

Sterilising the seed means putting a pinch of seed into a test tube together with a small amount of dilute bleach and a drop of washing-up liquid, resealing the tube with a rubber bung, then spending ten minutes (some species with thick seed coats require much longer) slowly tipping the tube upside down, then the right way up, again and again, as the tiny seeds swirl and tumble in the sterilising liquid like dark specks in a small cylindrical snow globe without a mini model Santa inside. Even without a Santa, watching the seeds whirl and spin in the solution, testas fading from brown to pale yellow, is, in its own way, mesmerising. After all, each spinning fleck could be a future orchid.

On this first occasion, as I watched the Bee seed tumbling around in the test tube, I experienced a hubristic rush. My plan for widespread orchid rewilding had begun! I set the test tube down. The seeds within settled. Very soon (or so I thought) Bee protocorms would be forming on my modified Malmgren's Medium.

It didn't work out that way.

First off, when I released the pressure cooker's valve, let it cool, then opened the heavy lid, I noticed my carefully prepared petri dishes full of Malmgren's Medium had transformed into dripping, misshapen plastic pancakes. I had put the dishes into the pressure cooker with their lids on. The pressure had crushed

The orchid outlaw's kit: a nondescript bag for transporting plants and earth; a trowel for digging and planting; an Ordnance Survey map (phone signal is seldom reliable when searching remote locations).

Referred to as 'Lady traces' by William Turner in 1548, Autumn Lady's-tresses is one of the first orchids described in English. Centuries later this species helped Charles Darwin explore the co-evolution of flowering plants and insects.

The Bee Orchid not only evolved to look like a female bee sitting on a pink flower, it feels like a bee and emits over a hundred different chemicals to smell like one too.

In an era when the supposed aphrodisiac properties of their roots mattered far more than their blooms, Gerard's *Herball* divided orchids into families of 'stone' (Elizabethan slang for 'testicle'). The accompanying woodcuts emphasise the plants' testicle-shaped tubers.

Of Dogs stones. Chap.98.

✳ *The kindes.*

Stones or Testicles, as *Dioscorides* saith, are of two sorts, one named *Cynosorchis* or Dogs stones, the other *Orchis Serapias*, or Serapias his stones. But bicause there be many and sundrie other sorts differing one from another, I see not how they may be contained vnder these two kinds onely: therefore I haue thought good to deuide them as followeth. The first kinde we haue named *Cynosorchis* or Dogs stones: the second, *Testiculus Morionis*, or Fooles stones: the third, *Tragorchis*, or Goates stones: the fourth, *Orchis Serapias*, or Serapias stones: the fift, *Testiculus odoratus*, or sweete smelling stones, or after *Cordus*, *Testiculus Pumilio*, or Dwarffe stones.

1 *Cynosorchis maior.*
Great Dogs stones.

2 *Cynosorchis maior altera.*
White Dogs stones.

✳ *The description.*

1   Great Dogs stones hath foure, and sometimes fiue great broad thicke leaues, somewhat like those of the garden Lillie, but smaller. The stalke riseth vp two hands high : at the top whereof doth grow a great thicke tuft of carnation or horse-flesh coloured flowers, thicke and close thrust together, made of many small flowers spotted with purple spots, in shape like to an open hood or helmet. And from the hollow place there hangeth foorth a certaine ragged Chiue or taffell, in shape like to a foure footed beast. The rootes be round like vnto the stones of a dog, or two Oliue berries, one hanging somewhat shorter than the other, whereof the highest or vpmost is the smaller, but fuller and harder. The lowermost is the greatest, lightest, and most wrinckled or shriueled, not good for any thing.

2   White Dogs stones hath likewise smooth, long, and broad leaues, but lesser and narrower than those of the first kind. The stalke is a span long, set with fiue or sixe leaues clasping or embracing the

Not all British orchids are called 'orchid'. The flowers of the Marsh Helleborine are as attractive as those of exotic tropical species. Habitat loss and pollution mean, like all orchids, Britain's helleborines are suffering unsustainable population decline.

The reasonably widespread Common Spotted Orchid produces spires of speckled pink-and-white blooms. Its genus, *Dactylorhiza*, means 'finger root', so-named because early herbalists likened the shape of their tubers to a human hand.

The tubers of the Early Purple Orchid were once sought-after as the main ingredient in the popular hot drink, salep. Although its population is declining, it remains quite widespread and can be seen on hedge-banks and verges.

Sometimes found in large colonies on chalky south-coast cliff tops, the blooms of the Early Spider Orchid herald orchid flowering season in Britain. A kind of Bee Orchid, it dupes pollinators into thinking it is an insect.

Germinating orchids need to experience winter – also known as some months in the fridge.

The suspicious powder on the digital scales is agar, an important ingredient in orchid propagation.

Each one of these specks is an orchid seed.

Common Spotted Orchid tubers grown from seed ready for guerrilla planting.

A 'little man' orchid, the Burnt Orchid has flowers shaped like diminutive people. In less than a century this species has suffered a huge decline across Britain. Despite these losses, the law offers it no special protection.

Variously compared to billy-goats' beards, tongues and lizards' tails, the Lizard Orchid has one of the strangest-looking flowers of any British wild flower. This is one of a few native species which may benefit from climate change.

With flowers looking less like butterflies than flying swans or angels, the Greater Butterfly Orchid comes into its own by moonlight, when the flowers seem to glow and its lily-like fragrance attracts large moth pollinators.

This small pot holds a salvaged fragment of land including orchids, microfauna and microorganisms. For millennia that kind of diversity flourished on these islands but, as this picture shows, today our land's living heritage is too often destroyed.

them into hardened nuggets of modernist sculpture. They weren't designed to withstand that much heat. Besides, I needn't have sterilised them. They came in a sterile bag. I should have simply opened them inside my air-purifier box and poured in the sterile medium. Instead, they had imploded.

Fortunately, I still had half the pack of petri dishes left in the bag – but I'd left the bag flapping open outside my sterile plastic box on the kitchen table. They would be contaminated. How contaminated? I didn't know. I grabbed the bag and put it in the box. Because the bag had been sitting in an invisible rain of microbes, putting it inside the box contaminated the whole work area. I grabbed a sponge dripping with dilute bleach, wiped the box down, took a petri dish out of the bag, lifted off the lid, grabbed the jug of Malmgren's solution (fortunately there was still some left) and tried to pour it in. As days in October often are, the ambient temperature was cool. The agar in the medium had solidified. It wouldn't pour.

Cursing, I whacked the jug back in the microwave for a few minutes, praying that would be enough to kill off any micro-organisms hitching a lift in the medium or on the jug and to soften the agar. Meanwhile, this delay meant the Bee seeds were drifting around in the bleach solution for longer than they should. Having them in bleach for too long could kill the embryos.

The microwave whirring, the solution in the jug recooking, I quickly stabbed the needle of a syringe into the bung of the test tube containing the seeds and drew off as much of the dilute bleach as I could, leaving a moist clump of pale seeds at the bottom. With a different syringe, I squirted a few millilitres of distilled, sterile water back in and gently shook the tube, to rinse the seed, then drew that water off and squirted in some more to rinse again.

In the meantime the microwave was pinging. I removed the Malmgren's, carried it to the plastic box (not a great idea as, once again, microbes in the air could fall into the black soup and make it home – I really was losing that game of chess) and carefully poured half a centimetre's depth of solution into the hopefully uncontaminated petri dishes. Something suspicious had happened – the black solution was gloopy. The agar hadn't spread evenly through the medium. It needed reblending. I had no time for that. I replaced the lid of each dish and watched them immediately steam up from the warmth of the solution. I had to wait for that to cool before I could sow the seed.

Waiting gave me time to contemplate what I thought I was doing. I accepted it was a first attempt, a learning experience, so I pressed on, squirted a final amount of distilled water into the test tube with the seeds, tugged off the rubber bung, swirled the water and seed, then sloshed the smallest amount I could into the nearest petri dish, moving the dish around so the water carrying the seeds flowed across the surface of the lumpy black jelly, spreading seeds with it. There were far too many seeds in that dish. Oh well. I squirted more water into the test tube, swirled it again and splashed most of the remaining seeds onto the next petri dish, then put the lids on.

I had used all the seed, hundreds of seeds, in two petri dishes, the majority of them into the first dish. If those in the first dish ever germinated, the young orchids would be so entangled they would be very difficult to remove. I had about ten more petri dishes, now not sterile. Behind my face mask, despite the cold, I was sweating. I was teetering on the verge of defeat and the battle was not over. Lying on the worktop were some pre-prepared strips of Parafilm. Parafilm is clear tape used to seal gaps in sterile dishes, like the join between the lid and the dish of a petri dish, and to hold the lid in place. It works by a kind of surface tension, but pull it too tight, too fast or at slightly the wrong angle and the Parafilm will snap; not tight enough or too slow and it will not stick to the dish and will not prevent microorganisms from entering. It's infuriating stuff. That day the Parafilm snapped many times.

★

Those first petri dishes were not a success. One developed a thin, fluffy white skin. The orchid embryos died. The other grew a cloud of green fungus with the same devastating outcome. Putting them on a shelf, wrapped in foil (seeds have to germinate in darkness), only to unwrap them with such anticipation after a week or two and find contamination like that is to witness carnage on a tiny scale, all the more tragic because it emerged from my attempt to help those orchids grow. Instead, I had bred some kind of bacterium and a fungus, but that was not my intention. That was not the most disappointing element of what I had caused. The chances of orchid seed germinating in the wild are very slim, but in my bumbling attempts to aid them I had stripped away even that chance.

On that first occasion there had been multiple opportunities for contamination, but in subsequent weeks, months and years, as my experience increased, I limited the numbers of petri dishes growing unwanted fungi or bacteria. It still happens, though, and often I'm unable to fathom where the contamination comes from. A single bacterium, particle or spore is all it takes for those tiny tragedies to unfold. Who knows, perhaps the fungus or bacterium growing on that jelly is incredibly rare. I wouldn't know. I don't think any human does, but they are one of the many ways orchid seeds can meet an untimely demise at the hands of a well-meaning amateur orchid propagator.

Other forms of death – or life – for those orchid embryos depend on how long they are treated in bleach, whether you soak them in water (and whether there is dilute sugar in that water) for days or months before sterilising; whether the seeds are exposed to light (and for how long) before you wrap each dish in foil to simulate the darkness of earth, label it with a marker (species, fungus, medium, date) and then put it into the fridge – or leave it out of the fridge for a few weeks, then put it in. For most, at some point, the fridge is crucial, because temperate orchids require the cooler temperatures of changing seasons. This *vernalisation* helps them develop, but it is not just about the fridge.

Over the years I learned that a slight change in one factor – light, temperature, time in bleach – can increase seed germination by 50 per cent or decrease it by the same, but it is not as simple as turning up the knob on one variable. Temperature, light, seed-coat thickness, length of time being bleached, time in and out of fridge and darkness – they complement and counteract each other. If anything demonstrated how much of a bumbling ape I was, this was it: chasing the sweet spot of orchid germination by attempting to unravel and recombine a few threads of a complex web which nature balances sufficiently accurately much of the time. Fortunately, as each orchid seed capsule contained thousands of seeds I had plenty of embryos to experiment with. Some of those experiments worked.

After weeks or months wrapped in foil on a shelf, the moment of reckoning arrives. Carefully peeling back the foil reveals whether seeds have become protocorms – pale tiny tubers or little fuzzy blobs a few millimetres long extending filaments into the agar around them. Sometimes round, sometimes pointed, orchid protocorms can be as varied in appearance as the mature plants. The first time I saw tiny Bee and Common Spotted protocorms growing in petri dishes I had prepared, I was astonished. It had, despite everything, worked! In those dishes, because of me, grew a new orchid generation.

When the protocorms grow big enough, with a few species (particularly Marsh Orchids) it's possible to plant them out in suitable soil, like planting peas. However, after spending months in a moist, sterile petri dish, the chances of those planted-out protocorms falling foul of pathogens or drying out are high. For this reason, another trip to the sterilised kitchen and air-purifier box (and more risk of contamination) is advised. On this visit the plump protocorms are *replated*. This process involves using plastic forceps to lift them out of the medium in that first petri dish and place them onto fresh medium in a bigger container, with fresh fungus, more space to grow and more oxygen to breathe.

At this point some species need to go into weak light and cool temperatures (I put them under a plastic-roofed lean-to that never got direct sun at the back of the house); for other species, it takes another season growing in the fridge before they can be planted out into pots of orchid-friendly earth, sand and grit. Some like to stay in the fridge until the protocorms produce their first leaf; some need to come out of the fridge for some months, then go back in. The important thing, as I learned after much trial and error, is to try to mimic nature. Despite all our available technology, that proved very difficult for a human to do.

Eighteen months after my first attempt with the Bee seeds my micropropagation technique had improved. Eighteen months is not long in terms of orchid growth – but it is a fair bit of time in which to learn. I had probably made every rookie mistake it is possible to make in a home-made lab. I forgot to mix the soup properly before pouring it into the petri dishes, so the growing medium in some dishes stayed liquid while in others it was as hard as plastic. I accidentally spilled seeds at the critical moment. I splashed bleach over my favourite T-shirt, forgot to sterilise my gloves before putting them into the propagation box and miscalculated the number of petri dishes I needed (so had left-over seed with nowhere to plant it). My medium bubbled over in the microwave; I melted dishes in the pressure cooker. I forgot to add critical ingredients to the medium, I mislabelled things, but I also learned that simple kitchen cling film can work just as well as capricious Parafilm.

Every mistake I made, I vowed never to make again. Which, as it turned out, wasn't always the case. Nevertheless, after those eighteen months I had parcels of petri dishes wrapped in foil filling the fridge's salad drawer; bigger plastic containers housing young plants took up another shelf and a half. My first twenty

Common Spotted Orchids had made it through the process alive and were growing in plastic pots sunk into a bigger tray of sand in the yard (the sand kept the pots moist and the temperature of the soil reasonably constant). I had some success with the Bees too; their tiny tubers were waiting to be planted into pots full of earth dug from Dad's lawn. Now I knew the importance of mycorrhizae, I was careful to make sure I had the right soil containing the right fungi gathered from woods, dune slacks and meadows to grow young orchids in.

It seemed I was improving my tactics, but it was a game that would last years. A single successful petri dish, then a successful replating, then successful planting out into pots led to dozens of plants waiting for deployment into the world. Hundreds more protocorms were in the fridge. Not all of them would make it alive but, finally, I was not just on the verge of rescuing a few colonies of wild orchids from diggers; I was on the verge of actively boosting their numbers. Except my plan had a flaw: the seeds I had were not the rarest. I wasn't getting the whole range of British orchid seed flowing into my kitchen lab.

True, not all native orchids (the Bog Orchid, Lesser Twayblade, Bird's-nest, Ghost, Coralroot) are suited to urban locations and lab propagation. I had no intention of trying to reintroduce *them*. However, I did have ambitions to branch out to rarer native species for which I could find suitable homes. If I could propagate and reintroduce these species, an additional dozen plants in the country would be a significant percentage gain. If they were encountered unexpectedly by someone at the edge of, say, a supermarket car park, that could attract local, maybe even national, news. That might raise awareness of Britain's orchids, their forgotten role in history, the plight they were facing, the plight Britain's wildlife was facing, and that might mobilise people to do more to protect it.

With this goal in mind, I had my sights set on a handful of rare species. I was particularly interested in the Lizard, simply because it was such an arresting and bizarre plant, and the Burnt

Orchid, because it was a minute beauty, suffering one of the steepest declines of any wild flower. They were also a species that had once grown in my part of England but were now extinct. I wanted to reintroduce them. Obtaining seed from these plants, which grow in so few locations, would not be easy; besides, the Lizard is Schedule 8. Collecting its seed would involve venturing deeper into outlaw territory.

A NOSEFUL OF LIZARD

To my mind the Lizard has the most extraterrestrial-looking flowers of any British plant. Their flower spikes can be metre-tall masts garlanded with multiple spindly tongues. Bone-white, blotched with alien, pink markings, darkening to claret towards their twisted, tangled tips, each long labellum emerges from an open mouth formed by a green-striped hood. These elongated labella look like tongues to me; others, such as John Gerard, compared them to the tails of small lizards racing up the flower spike (Gerard described Lizard flowers as 'in forme like unto a lizard, because of the twisted or writhen tailes and spotted heads').[1]

Although the lizard comparison gives this species today's name, in Gerard's day they were more often associated with goats. He referred to them as 'Goat Stones' and '*Tragorchis*', meaning 'goat orchid' (from the Greek *tragos*, 'he-goat'). Fifty years after Gerard, Parkinson's *Theatrum Botanicum* explained why: 'These *Orchies* are so named not onely because they have a strong foule sent like a Goate, but that most of them have long tailes like beards hanging downe from them.'[2] The 'foule' scent Parkinson described adds to their strangeness. For Gerard 'the stinking and lothsome smell and savour they are possessed with' was the main reason why the Goat Stone was 'seldome or never used in phisick'.[3] The current scientific name retains this caprine association:

Himantoglossum (Greek for 'strap-tongued') *hircinum* (Latin for 'goat-like').

I've not spent sufficient time around goats to tell whether the Lizard smells like one. That was why, when I first encountered them, standing proud beneath a clear June sky, their bizarre tongue-like, tail-like, beard-like, strap-like petals jostling in a warm breeze sweeping off the Bristol Channel, I approached them with caution. I intended to make up my own mind about their scent. Fearing the worst, I stuck my nose into a tangled mass of petals and took a tentative whiff.

I expected a stench of slaughterhouses and the nether regions of beasts. Instead, there was a mawkish hint of honey and vanilla. Unexpected. So too was an undertone of decay. It conjured dim memories of a damp white rat I had dissected in a biology class. The rat had been preserved in embalming fluid. *That* was the perfume of the Lizard: embalmed rats and vanilla.

I was fortunate to find those Lizards on that hot day: their captivating flowers are entirely absent from most of Britain. Darwin's description of them as 'extremely rare' suggests they were perhaps even rarer in the mid-nineteenth century.[4] Records show Lizards once spread across southern and eastern England and north into Yorkshire. Since those days the bizarre, stinky Lizard's territory has shrunk by over 82 per cent.[5] Despite this loss, I hadn't heard of any officially sanctioned plans to raise Lizards from seed and reintroduce them to suitable habitats. Even if I had, I doubt it would have dissuaded me from doing what I could to try to reintroduce it to areas where it hadn't been seen for decades, if at all.[6] To this end, I was returning to where I had, years earlier, first taken a noseful of Lizard. This involved striding down a public footpath next to a small supermarket towards a gate covered in warning signs.

The signs cautioned Joe Public not to deviate from the right of way and to beware flying golf balls. Needless to say, in the past I had ignored that advice and roamed the fairways beyond, avoiding golfers while I photographed Bee, Pyramidal and Lizard

Orchids in the knee-deep rough. On this particular occasion, though, I hoped to have timed my visit to miss all those fabulous flowers. I was there to find Lizard seed pods.

The colony of Lizards made their home in a dip surrounded by tussocky dunes caught between two fairways. I hurried across the worn grass past four men towing trolleys. Buoyed by the anticipation of embarking on this phase of my clandestine programme of propagating one of Britain's rarest orchids, I greeted them cheerfully. They stared as if my greeting had been a veiled insult, then strode on, muttering together, probably about how inappropriate it was to allow non-trolley-towing weirdos access to the fairway. I didn't mind. A superior satisfaction overtook me. I had already spied the rickety skeletons of spent Lizard spikes, the dried remnants of their 'writhen' splendour clinging to the tips of seed pods. It was disappointing that those golfers showed zero interest in the rarity of what grew around them. I could have trampled those Lizards, uprooted them, flung them around, and (I guessed) the only thing those golfers would have cared about was whether I had moved one of their precious little balls. The feeling of superiority segued into sadness. How many others shared their attitude?

I teased a folded sheet of aluminium foil out of my pocket – it's the best temporary transportation device for orchid seed pods. After checking no one was watching, I turned my attention to the nearest withered flower spike. It was lined with plump brown seed capsules. Some had already split. I gave the spike an experimental tap. Brown smoke wafted away. Probably too far gone. There was no point taking that one if the pods were already empty. I needed another, not quite so dry.

I scanned the bleached vegetation for withered Lizards. A few loomed here and there deeper in the dip. As I descended towards them I had the sensation of being watched. I looked up.

A man was on the path. Against the blazing sun I made out dark clothes, an athletic build, an absence of putters and drivers, a baseball cap and a face angled my way. Was he *watching* me? My heart plummeted.

Casually, I slipped the blindingly conspicuous silver foil back into my pocket and, without so much as a passing glance at the Lizard spikes, I drifted into the dip, then up the other side. It was my best impression of being a slightly lost wanderer, nothing more. If challenged, I would plead ignorance.

From the rise, I gazed wistfully back as if soaking up the vista of dunes, little flags and golfers, but out of the corner of my eye my attention was on the watcher. He was still watching. Now he was holding a mobile phone or walkie-talkie to his face. Golf-course security? A protective local who knew the value of those plants?

I hadn't done anything indictable, but as long as he was there it was too risky to harvest seed. Coming back another day would be risky too – from the parched appearance of the seed pods, by the following day they might all have split in the heat and it would be another year before I had any chance of collecting more.

Nonchalantly I returned to the path and headed away from the watcher, towards the beach. A hundred metres on, the path became a boardwalk between bulrushes. I paused there to look back along the path and . . . he *had* followed me. When I looked, he paused, but he was there, talking, watching. Unnerving.

I picked up my pace through the bulrushes into dunes at the edge of a wide deserted beach. I headed right and, as soon as I was out of sight, slipped into the dunes. Concealed, I waited. Moments later the man appeared. He jogged uncertainly towards the distant waves, looking left and right, searching. As he walked forward, his attention elsewhere, I doubled quickly back along the boardwalk. Whoever he was, whatever he wanted, I had no intention of finding out. The price could be too high. My steps muffled by windblown sand, I raced to where the Lizard pods were waiting.

12

BURNT-TIPS AND LONGHORNS

The Burnt (or Burnt-tip) Orchid, *Neotinea ustulata*, has not grown wild in Devon for close to a hundred years.[1] Sheathed in pale-green leaves, the flowers of these diminutive orchids first appear as tightly packed, burgundy-coloured, grape-like buds. The flowers open sequentially up the spike, revealing mostly white flowers with pale pink spots, so the flower spikes assume a two-toned appearance: where the flowers are open the lower part is white; the upper tip, where the unopened buds remain, is deep red, lending it a scorched appearance. This is the origin of their common and scientific names: *ustulata* means 'scorched-looking'. To me, they look less scorched than like black-cherry-sauce-dipped gourmet desserts, each fashioned out of layered frills of white chocolate, each frill crafted to resemble a tiny person, each 'person' randomly speckled with raspberry coulis like the lesions of a delicious pox. Why do so many of Britain's orchids, aside from the Burnt (including the Military, Lady, Monkey, Man, Dense-flowered), have this humanoid shape? No one knows. What interests me is bringing these little beauties back to the landscape . . . if I can.

Burnt Orchids were first written about in English in 1633 in Thomas Johnson's revised edition of Gerard's *Herball*. An apothecary who plied his trade near the Barbican in London, Johnson appears to have been blessed with inordinate amounts of energy and skill, for, aside from his work as apothecary and physician, he expanded, revised and translated multiple herbal compendiums as well as writing his own. He was the first person to sell bananas in England, wrote the first account of an ascent of Mount Snowdon (in 1639) and embarked on several journeys around England and Wales, finding time to draw and describe the plants he found.

Until then the most comprehensive listing of British plants, Johnson made Gerard's *Herball* even more so. In it he notes that the Burnt Orchid 'groweth in many hilly places of Austria, Germany, and England'.[2] As with most British orchids, its medical purpose lay in its ability to stir 'lusts of the flesh'. When Civil War divided England, Johnson set aside his career and his botanical writing, enlisted in the King's army and became notorious for dashing acts of bravery, until the moment in 1644 when a bullet entered one of his shoulders and 'the best herbalist of his age' succumbed to fever. His burial place is unknown.[3]

In Johnson's time, the Burnt Orchid may indeed have flourished in 'many hilly places', but it has since suffered a catastrophic decline. Once recorded in over 350 sites in Britain, by the late 1990s it existed in only seventy-five, some home to only a couple of plants. It was last recorded in South Devon in 1932, in North Somerset in 1961, in Surrey in 1966 and in South Yorkshire in 1973. In Oxfordshire it was last seen in the wild in 1982. I could go on.[4]

This small, exquisite orchid retreated, for the most part, to a few chalk downs in southern England. Over the past century many of these colonies have disappeared too. The land has been ploughed, over-fertilised, doused with herbicides and pesticides. Pollinator numbers have been decimated, the chemical and biological composition of soil has changed and the mycorrhizal fungi these orchids need have been badly affected, probably rendered extinct in places, but no one knows. For these reasons, for most people, it now takes a special pilgrimage to find Burnt Orchids. Thankfully, so far, they have clung on in places where orchids are afforded some protection. Some are on private land. Some are managed closely by Natural England. Some have geographical defences (distance), others man-made barriers (barbed-wire fences) or administrative hurdles (permission to access, forms to complete, risk assessments to sign). All of these were in place when I set out one day to look for their seeds. There was also an unexpected challenge.

I spotted the herd of English longhorn cattle some way off across the chalk down where the land rolled away on gentle, butterfly-flickering slopes mottled with low grass and the dappled palette of wild thyme, azure milkwort, cadmium-yellow bonnets of horseshoe vetch and bird's-foot trefoil and the blue floating suns of devil's-bit scabious. As I descended the slope it was not a habitat that offered much cover.

On my previous visit I had found the raspberry-ripple blobs of the Burnt Orchid scattered across the pasture. Since then things had changed. There were a lot more fences. And, for some reason, the stewards of the SSSI had introduced those longhorns, despite the fact that large herds of cattle are not always the best company for dainty orchids.

English longhorns are far from demure dairy cows. They appear to be direct descendants of Tur, the Slavic bull god who shook his horns to create earthquakes. Their horns are mythical, each as long as my arm. As I approached, the great beasts stopped grazing and glowered. A few flicked sloppy tongues into bovine nostrils. They stared and huffed, grumbled, grunted and jostled each other's fly-buzzed flanks. They were about 30 metres away. I judged it safe enough. I was pretty sure they would amble off. I climbed the fence and dropped into their paddock.

Google will inform you that, hostile appearance aside, English longhorn cattle are 'usually friendly'. 'Usually' was not the re-assurance I needed right then with forty pairs of eyes and forty sets of horns focused intently on me. The fence behind me offered an easy exit. Fifty metres ahead, however, stood the fence beyond which, if I remembered correctly, I would find the Burnt Orchids and potentially the seed I was seeking for the benefit (I told myself) of the country. If I could get over either fence, I would be safe, but I wasn't sure if my chances of winning a foot race against charging longhorns were good enough to get me to where those orchids grew.

Warm and close in that way breezeless grassland has, the air carried the thrum of the busy A-road a mile away, where passengers sat oblivious to the drama unfolding on that down. They were probably also oblivious to the huge Burnt Orchid colony, the biggest in northern Europe, growing on the other side of that field.

The sound of cows slurping and snorting joined that of the distant traffic, the tumbling liquid notes of skylarks, tick-tocking insects in the grass like a million diminutive clocks counting down to . . . the moment that bumping, butting wall of horn and muscle decided they didn't want a gangly bipedal intruder. Time to evict it. Or perform the herd's version of an experiment: How fast can we make Ape run? What does it feel like to trample Ape?

Eyes bloodshot, they shuffled, grunted, edged closer, two, three, four deep.

I had not moved. I was calculating. Would they lose interest? How close could I let them get before I beat a hasty retreat? If I strode towards them, would they part obligingly? Who was I kidding? I imagined the headlines: 'Cow Kebabs Crazy Orchid Collector', 'Orchid Loon Longhorned'.

Has anyone studied the degree to which cattle sense fear and how they respond? Perhaps they smelled panic oozing from my pores. Perhaps it was perceptible in my posture. Whatever – however – the dawning awareness that they could now overturn centuries of obedience to man sparked in their big eyes. Each weighing several times what I do, those great horned beasts shuffled closer. Panic churned my gut. I tried to control it.

The cows fixed me with mutinous glares. I was suddenly very

aware that, compared to those big animals, each honed to power all that weight into their horns, I was weak. Slow. Fragile. A kind of paralysis invaded my limbs. I tried to fight it. Then something stirred, a new instinct, new but weirdly primal – something so deeply embedded and until then overlooked that I had no inkling it was there, gathered inside me. Suddenly called forth by circumstance, the wisdom of a thousand generations of nomads and farmers surged through me, threw up my arms and tore a primordial roar from my lungs.

The Ape cried out, peculiar, tall, fearless. The herd leaders faltered.

Ape roared again.

Rearing, the cows twisted away. They stumbled into others. The herd bolted, wheeling across the pasture, a hundred and sixty heavy hooves drawing muffled thunder from the turf. It was like the beating of drums, an exhilarating rhythm, a summons to ancient rites unheard for centuries.

I stared at their retreat, as shocked as they were. In that instant I had been Master of Cows, a conduit of some profound figment of a distant epoch when man was domesticator, herdsman, protector of the living land.

In the far distance, they swung around, further and further, until they were heading back towards me, a barrage of hooves and horns. Heads high, the great beasts charged.

Could I trust them to obey again? I doubted it. The fluttering fear I had beaten down flew free. As usual, the orchids' siren song coaxed me into craziness. Together with my firm belief that the moral right – the 'moral obligation' – was on my side, I desperately launched myself towards the distant fence beyond which Burnt Orchid seeds might be standing in the heat-yellowed grass.

The cows were closing in. Angry. I urged my reluctant legs to rediscover lost youth, to hammer the grass, to drive me into a future where I might save orchids for the nation. My bag was heavy. My boots were lead. The air was syrup. The fence remained stubbornly distant. My laboured wheezing drowned the sound of pursuing cattle. I wanted to look back, to calculate my chances, but I feared the second my attention was not on the ground it would open up into a badger sett. I would plummet in, break an ankle, then have horns and hooves all over me.

Long seconds later, I slammed into the fence and dragged myself gracelessly over, grateful that of all the fences I had encountered that day, that one was not topped with barbs. My bag smacked my head hard. I had lost my sunglasses, but I was safe. Safe, in the vicinity of Burnt Orchid seed heads rising from chalkland sward like diminutive brown baby's rattles.

That was just one of the missions I embarked on during those early (and not so early) years. Over that time I established a reasonably successful calendar of Orchid Outlawry. Late spring and mid-summer saw me engage in a flurry of orchid-related activity to make the most of that short window when the majority of Britain's orchids flower. In July and August I pursued seed pods and planted some species in petri dishes; in October I planted other seeds in petri dishes and guerrilla-planted tubers and rhizomes. Over winter I trawled planning applications, then the year began again. I did all this from a house which, from the street, looked like any other two-up-two-down terraced house, product of an era when British people became the first in the world to let the factory shift and train timetable dictate the pace of life, a time when citizens in search of employment swapped small, rural, agricultural settlements for crowded cities.

To those newly arrived country folk of the nineteenth century, uniform terraced houses like mine may have had a certain appeal, for each little house had its own outdoor space. So important was this space that the Model Byelaws, arising from the 1875 Public Health Act, enshrined its size in law: a minimum of 150 square feet (14 square metres), not including privies.[5] Small as this was, these yards offered a scrap of the outdoors where some attempted to raise livestock and poultry or even grow a few edible or medicinal plants. In this sense, terraced houses offered a stepping stone between farmland and tarmacadam, a gesture to the way of life of pretty much every human generation that had gone before. My 150 square feet of brick-bound space is probably unique. What sets it apart is not so much the plant pots, although there are around fifty of various sizes, but what those pots contain.

In winter the pots appear unassuming, the yard bleak, abandoned. Some pots sport low crowns of weeds, others hunker down beneath panes of glass. In some, haloes of leaves form, hugging the soil surface; green nubs poke their noses out in others. At that time of year these are the only clues that this yard serves as the hub of my guerrilla-rewilding missions. In March things start to change. Green shoots inch towards the sky, extending upwards like telescopes or unfurling, like tulips, one leaf at a time. First, from nests of dark-green leaves splashed with deep black-burgundy patches, the flower spikes of Early Purple Orchids push upwards. One by one, the flowers open. They look a bit like flamboyant mauve hyacinths. Past generations saw their leaf-markings as the stains left by Christ's blood falling on them as he bled on the cross. Hence one of the common names of this orchid: Flower of Gethsemane.

Around the same time, the buds of a few stumpy plants with small lime-green leaves open to reveal fuzzy brown flowers graffitied with pale scratches often forming a rough letter H. A corona of five light-green 'legs' surround these strange little ciphers. Centuries ago, someone thought the blooms looked sufficiently like a little fat-bellied spider to call it the Early Spider Orchid. The name has stuck ever since. Next, the blooms of the Green-winged Orchid arrive, like huddles of dainty green-striped bonnets. From mid-May to mid-July, the yard explodes with pink, white, yellow, light-green and vermilion pyramids, pillars and columns, swaying bouquets (Musk, Common Spotted, Common Twayblade, Heath Spotted, Southern Marsh, Early Marsh, Heath Fragrant), interspersed between ethereal green-yellow ghosts (Greater and Lesser Butterfly) and alien-like twisting tails (Lizard). The clownish hairy faces of flowers evolved over millions of years to resemble insects (Bee) and small clouds of flowers shaped like diminutive cartoon people with stubby limbs and hoods (Military, Monkey, Lady, Burnt) appear here, many of them highly endangered, most of them suffering massive population loss in the wild.

As summer progresses, the flowers wane and many of those blooms become capsules stuffed with maturing seed. Summer segues into autumn and the last of Britain's wild orchids, Autumn Lady's-tresses, raise their little spires hung with pale bells, then, as early winter gales strip the last leaves of the neighbourhood's trees, they die down as the next generation of the species that flowered earlier in the year are already forming leaf rosettes awaiting next year's spring to start the process of flowering and seed-setting again.

It is a cycle of growth and rebirth, and the creation of new generations which these plants and their ancestors have engaged in for many millions of years, but as human activity disrupts that cycle, here they are, in this yard constructed to cater for human migrants from the countryside, but now turned into a halfway house, a nursery, a barracks, a refugee centre, a seed orchard. That is what makes this yard unique.

With those pots as my seed orchard and with seedlings growing, with my kitchen lab and my pots of plants, I was doing what well-funded national institutions were seemingly not. And I was not overseeing taxpayer-funded projects that failed to prevent on-going species loss.[6] I was not debating definitions of Endangered and Critically Endangered or mulling over whether a species should be downgraded to Least Concern. To me such categories were counterproductive when every natural habitat and wild species in the country was – is – at risk. To me there were more pressing concerns than redrawing definitions or meticulously observing biodiversity crash.

Time and money spent plotting their genomes was not time spent in the field protecting them. Yet, judging by the numbers of scientific papers on orchids being published, for decades that was what the vast majority of scientists were doing. Over the same period government policies and environmental legislation had failed to stop species decline. There was no consistent, robust, effective legal or political action to save Britain's disappearing orchids, which, in turn, was a damning indictment of Britain's defence of its nature.

It seemed to me that I was paying more attention to the needs of these rare plants than the government or the law. All I wanted to do was take what Darwin and Malmgren and most of all orchids had taught me to bring those complex, curious, endangered plants back to the people, people who probably had never seen one, never knew they existed on these shores; back to places where they had not been seen for more than a human generation. That would make me proud. What I could not do, however, was tell anyone. I already had a full-time job. What I was attempting to do was a lot to take on alone. I needed a partner in crime.

13

ENVIRONMENTAL OUTLAWS

A police siren chirps to life, then howls. My attention snaps to the rear-view mirror. Pulsing blue lights. A squad car bearing down. My heart lurches. Instinctively, I check the speedometer. It's within tolerable vicinity of the limit.

'Cops,' I say.

Sylvia, my wife, casts a look behind.

'They just want to get past.'

She's probably right.

I indicate, pull over on the rural lane, wait for the police car to sweep by.

A decade has passed since I first encountered those Bee Orchids in my parents' garden. For years now, those creatures – Bee, Military, Monkey, Lizard, Burnt, Common Twayblade, Common Spotted, Early Purple – have occupied my thoughts like troubled family members in need of safekeeping. I have spent many dawns rescuing and replanting orchids in the company of foxes; many evenings stooped over musky-clove-scented orchid seed; many weekends sowing seed on growing medium, replating, planting, watering, shading, guerrilla-planting. Over that time, orchids have reshaped how I see the world. Now, wherever I walk or drive, at least half an eye scans the green spaces I pass, assessing whether this spot or that might be a good site to plant or find orchids.

Orchids have also revealed to me not only a largely forgotten thread of our own culture's connections to them, but that all those green places are complex, living habitats, as are people, and slow-worms, orchids, earth . . . I know much more about Britain's orchids now and I am still learning. Well, *we* are. This is not a lone venture any more. Before we married, I had to tell Sylvia what I got up to in the early hours of Sundays.

This confession came one evening during a David Attenborough documentary featuring the rapid warming of the Arctic and the consequent deterioration of the health of polar bears and their newborn cubs. Sylvia was distraught.

'Climate change. Global warming. One day people'll be like those bears,' she exclaimed, 'losing weight, having children that won't survive. *Everyone* should be doing more to stop this.'

'I do what I can,' I said quietly.

'Recycling? Walking to work? Not eating meat? It's not enough.'

'I save orchids.'

She blinked.

'What?'

'Those edible mushrooms I said were in the fridge . . . the containers in foil . . . they aren't mushrooms.'

She frowned.

I opened the fridge, pulled out a foil-wrapped petri dish, one of the dozens stacked in the salad drawer, each labelled with its species and the date I'd put them in (so I knew when to take them out again).

We had only been going out a few months. I really liked Sylvia and had grappled with keeping this secret. Would she want to continue a relationship with someone who flagrantly broke the law? Would she report me to the police? Would the true identity of what grew in the yard end our relationship? Or bring us closer? Her reaction to the threat to polar bear cubs suggested the latter.

'Orchids?' Doubtfully, she squinted through the transparent plastic at the blobby protocorms in their sealed world.

'Baby orchids,' I explained. 'And all those pots out the back have got orchids in too. You can't see them at the moment, though. They're underground.'

'Underground?' Her frown deepened.

'Well . . . some aren't, some are.' I was nervous.

'Orchids.' She seemed to be considering everything anew – the fridge, the kitchen, the yard and me. Was she about to grab her coat and beat a retreat to a lifetime free of a crazy guy growing blobs he said were orchids in his fridge? 'Why are they in the fridge?'

'How much time do you have?'

As it happened, much to my relief, Sylvia had ample time for me to tell the whole story. More than that, she became a willing convert to the cause, less interested in the kitchen lab but more into anything online. Dexterously employing a smartphone she swipes through the virtual ether, monitoring local planning applications and environmental impact statements, cross-referencing them with databases of orchid sightings; checking for stories about developers and local councils destroying orchids; corresponding with 'contacts' in the orchid community (none of whom, by the way, know exactly what we do). From what she learns we decide where, how and when to act. That's our guerrilla orchid army. my wife, these orchids and me.

It'd be remiss not to mention our newest recruit, Nate. At his age our son is very much a trainee, but his arrival changed everything. Equipment from the kitchen lab had to be moved to another room, out of reach of curious little fingers. In a stroke of genius, one day I tried mixing one of the sterilising tablets we used for his baby bottles into the Malmgren's Medium to see if it reduced microbe contamination. It seemed to.

Most of all, though, Nate has become the reason we do what we do, because saving these orchids is about saving the future, and the future is more his than ours. With Nate's arrival, our mission gained a sense of urgency. We want him to grow up in a world bursting with the magic complexity of life, with healthy habitats and abundant ecosystems which will help his generation to thrive.

Now our kitchen is a war room. At first glance, it looks like any working family's kitchen. Last night's saucepans air-dry beside a Belfast sink. A school year planner is Blu-Tacked to the wall by the kitchen table. Beside term dates, grandparents' birthdays, parents' evenings and work trips, the weeks from March to October are highlighted in fluorescent pink or yellow and labelled: *E. Spider*, *Lady*, *Monkey*, *G. Butterfly*, *Lizard*, *Bee*, *Frog* . . . If you didn't know better, it might look like the ingredients of a witch's brew. Next to some names are alphanumeric codes: SU608745, ST490181 and so on. Those familiar with the Ordnance Survey maps of Great Britain will recognise grid references.[1] Combined, the names, highlighted weeks and grid references are reminders of where to go and when to gather seed from some of Britain's orchids and the locations of sites earmarked for development where orchids may need rescuing.

Open the fridge and you will find one shelf in the door stuffed with labelled envelopes. Each contains a vial of dried seeds. The labels match the calendar (*Lizard*, *Frog*, *Bee*). Some of these constitute our seed bank for germinating at a later date. Some are here to experience a cold spell before germination. The salad drawer, that awkward one you have to crouch to reach, is stacked with petri dishes and plastic tubs sealed with cling film and wrapped in kitchen foil. Take out a tub. Hold it upright. Unwrap it. Peer inside. Resting on a centimetre of charcoal-black jelly you will see a scattering of small, white, slightly hairy blobs: protocorms.

A couple of the kitchen cupboards are sagging on their hinges. Inside are stacks of plates, glasses, cups, tinned tomatoes and beans, bags of rice, packets of pasta, a large tin marked 'flour'. Pop the lid off that tin and you will find bags with labels displaying batch numbers. These are bags of agar powder, the very same which ends up in those petri dishes in the fridge. As you know, that process once happened on this kitchen table. For a while it migrated to the spare room up the narrow stairs – 'the Lab'. As Nate got older the Lab had to be packed away, its components stuffed into the understairs cupboard, ready to be reassembled when required.

Having full-time jobs and now a youngster has meant orchid time is limited. We continue to operate, though, without permits or authorisation. As our only qualifications are a couple of science A-levels and we are not employed or supported by the Royal Botanical Gardens of Kew, Natural England, the National Trust, the World Wildlife Fund, the local Wildlife Trust or any other recognised institution, our requests for information and access to sites are almost never answered, let alone approved.

It's frustrating, but I don't hold it against the decision-makers. To the national and international conservation community we are nobodies. But as nobodies operating on the other side of a legal boundary, we have a freedom to explore avenues not open to those conforming to well-meaning (if flawed) legislation. Any institution would take some convincing that we are not irresponsible mavericks. Even if we succeeded in convincing them of that, technically we are still criminals.

That is why, as that police car bears down, I can't help thinking of the two of us as an eco-conscious Bonnie and Clyde. This is not too far from the truth. We dropped Nate with his grandparents for the morning and now a couple of dozen orchid tubers are sitting in plain view on the back seat in a lidless shoebox. We're on our way to plant them in a few carefully chosen sites: a grassy bank near a village church, the edge of a rural rugby pitch and a patch of ground near a war memorial which stands beside a crossroads amid high hedged lanes. They are all nicely visible places. As usual these days, I've also got a vial of seed mixed from several species, which, if I happen to come across a likely-looking location, I will scatter and let nature take its course.

The thing is, the strange and difficult thing is, now I have pulled over, the police car does not sweep past. It stops a couple of metres behind us, lights circling.

My palms are suddenly slick.

'What's going on?'

'They've stopped.'

Now I'm wishing I'd left the seed at home and covered the box. The orchid rhizomes and tubers are clearly visible on the back seat in a nest of damp moss. My gut sinks. To me that box screams, *Orchids being illegally transported!* Is this the moment when our save-the-future action costs us big? With the penalty for having these plants at £5,000 per plant and six months in prison, we're looking at, I guess, over £120,000 in fines. Will we have to sell the house? If we do, what will happen to the Guerrilla Orchid HQ? What will happen to Nate? Our jobs? How will we afford court fees?

Sylvia and I exchange a look.

A uniformed figure emerges from the squad car, saunters towards us.

I have the idea of pulling off my sweater and throwing it over the box, but by the time I've got one arm stuck halfway down a sleeve, I give up. It'll only look suspicious. I straighten out the sleeve, hold the wheel.

The uniformed figure is at the window. Heart hammering, I whirr it down. I'm flushed. I probably couldn't look more guilty than I do in that moment.

The officer stoops, peers in, surveys me, Sylvia, the interior of the car.

'Good morning, sir. Could you turn the engine off?'

I do what he says. He looks about my age. A bit world-weary.

Birdsong punctuates the silence. At the edge of the road is a verge. A breeze whispers through the window. I watch the grass stir and wonder if this is my last glimpse of freedom: a late October day, partially cloudy, a few yellow leaves. I find myself wondering if that verge would be a good place to plant orchids.

'Is this your car, sir?' British police can be very polite.

I unglue my tongue.

'Yes.'

'Do you know your rear fog light is on?'

'Sorry?'

I had expected, *Step out of the car, hands behind your head. I am placing you under arrest for removal of orchids without permission and on suspicion of illegally obtaining Schedule 8 species contrary to the Wildlife and Countryside Act 1981.*

'Your rear fog light is on.'

'Oh.'

'You didn't switch it on?'

'No . . .?' I look, dazed, at the dashboard. I'm not entirely sure where the rear fog lamp switch is. I'm not sure I've ever used it.

'Try . . .' He squints at the dashboard. 'That button.'

'OK.'

He walks to the back of the car, checks the light.

'That's done it.' He ambles back. 'You must've knocked it.'

He loiters, scans the outside of the car, the tyres . . . Am I meant to say something? I'm not sure. Maybe he's wondering how to bring up the awkward question of the rare plants on the back seat. Is he waiting for a bribe? I'm starting to wonder whether it's a ploy. Good cop–bad cop wrapped up in the same cop.

'Thanks,' I say again. 'I'll be more careful.'

He nods, slowly heads back to his car. He seems to be enjoying a short stroll in the sunshine.

'Is that it?' I ask Sylvia.

'I don't know.'

She looks out of the rear window.

'What's he doing?' I ask.

'Going back to the car.'

'Don't stare.'

'How am I supposed to know what he's doing without looking?'

'Is he going to bring us a formal warning?'

'If we drive off and he wants to, he'll just stop us again.'

'I'll . . . start the engine.'

We wait. The police officer does nothing. I indicate, check the mirrors, pull out using my best driving technique. That's the last we see of the policeman.

He said nothing about the nuggets of orchid life on the back seat. On the one hand, this was a massive relief. On the other, it illustrates how ineffectively the law protects our land, its life and the future. Having those orchids on display was like having a haul of cocaine in plain view and being waved on through . . . except, with a few telling differences. Cocaine is an addictive, controlled narcotic, a significant part of the £9.4 billion drugs market in England and Wales in 2020.[2] I don't think anyone keeps tabs on the annual global market in European terrestrial orchids. There's a reason for that: the market is very, very small – and orchids aren't addictive (at least not in a straightforward chemical way), which makes me wonder whether the authorities would care a great deal more about them if they were. Instead of investing millions of pounds in protecting endangered living nature for the future, hundreds of millions are spent every year on unsuccessfully policing an illegal plant product.[3] The illegal market for cocaine has only grown since the incredibly costly war on drugs began; in the meantime, largely due to legal activity, native species slip away.

Sadly, even if that fog light-savvy police officer had slapped handcuffs on us, in Britain, prosecutions for wildlife crime are rare and successful convictions rarer. 'The purpose of the relevant legislation', explains the Crown Prosecution Service (CPS), referring to wildlife crime, 'is to protect and conserve creatures and habitats that belong to all.' That sounds laudable, but given that even the government's own data shows that biodiversity and habitat loss have continued unabated for decades, the lack of prosecutions for wildlife crime don't seem to make sense. Statistics suggest a crime wave sweeping the land, stealing away those 'creatures and habitats that belong to all', yet the pitifully low numbers of successful convictions (annual convictions rarely reach double figures, while attempted prosecutions and reported crimes annually reach hundreds) suggest that this crime wave surges largely unopposed.[4]

One reason for this is that the CPS regards only specific wildlife crimes as 'in the public interest' to prosecute. These wildlife-crime priorities address areas that are 'posing the greatest current threat to either the conservation status of a species or which show the highest volume of crime'. For years it prioritised four areas: the persecution of badgers, birds of prey and bats; poaching (deer, fish and hare coursing); the disturbance of freshwater pearl mussel beds; and the illegal international trade in endangered species, primarily 'European eel, birds of prey, ivory, medicinal and health products, reptiles, rhino horn and timber'.[5]

In 2022, years after it should have appeared, 'cyber-enabled wildlife crime' was added to that priority list. Wild-flower theft or destruction is not specified as a priority. As of late 2021, no central record exists of wild-flower-crime prosecutions, reports or allegations of such crimes[6] – even though 19.9 per cent (around 277 species) of England's plants are at high risk of extinction (Critically Endangered, Endangered and Vulnerable).[7]

Then again, no central recording system exists for priority crimes (or alleged crimes) against bats, badgers or poaching either. Rare native reptiles and amphibians fare no better. Draining marshland or felling oak forests – unless they are already in a conservation area – is not considered a crime. Arguably of the greatest public concern are the trillions of beings existing in the soil, all of them crucial to nutrient cycling, carbon-dioxide sequestration and the health of plants and, therefore, insects and animals, including people, but the law makes no mention of them.

Knowing this makes me wonder who benefits from these laws. Over the period in which they have been in force, the alleged beneficiary, wildlife, doesn't seem to be benefiting a great deal. Are the main areas of CPS focus actually protecting our rarest species? It seems not. The CPS states that for any case it takes into account 'the biodiversity status of the species involved'. It appears to base its decisions around this on the lists of protected species contained in the Wildlife and Countryside Act 1981. The problem is, in over twenty years that list has not been adequately updated or corrected. Is the CPS aggressively pursuing crimes that damage Britain's natural world? Not really.[8]

As I discovered earlier in my orchid journey with those lost Autumn Lady's-tresses, there appear to be ways round laws, allowing environmental destruction if it benefits humans (no matter how short-sighted these benefits may seem and how much damage to the environment they may inflict). That is why, despite protests by conservationists, urban expansion continues and why £27 billion a year has been allocated to expanding Britain's road network by 4,000 miles between 2020 and 2025. It is why the HS2 train link between London and northern England is excused its destruction of rare habitats (directly impacting 108 areas of ancient woodland, thirty-three SSSIs, five sites of international importance and over 9,000 hectares worth of local wildlife sites).[9] It is why new oilfields in the North Sea can be tapped, the by-products of which will poison the planet for centuries – and why none of these are regarded as criminal enterprises.[10] How does any of this match the purpose of protecting and conserving 'creatures and habitats that belong to all'?

This leads to another question: the law implies that the 'creatures and habitats' of the land 'belong to all'. Who is this 'all'? All people on the planet? All Britons? While I agree with these sentiments, how many of this land's creatures and habitats really belong to 'us'?

Over half the land mass of these islands is owned by less than 0.06 per cent of its population.[11] This extremely undemocratic and unrepresentative percentage owns millions of rural acres, including grouse moors and stud farms which their owners have no qualms about using to claim millions of pounds in farming subsidies.[12] The land they own includes some of the 71 per cent of Britain dedicated to agriculture, which, almost by necessity (if its owner wants to claim subsidies), gets doused in pesticides and fertilisers, overstocked, polluted, drained and destroyed.[13] Only about 5 per cent of Britain belongs to millions of run-of-the-mill homeowners (like me, and, maybe, you).[14] In short, there are hundreds of thousands of acres of undeveloped land on this island – our island, our country – it's just in the pocket of someone else, most of it in the pocket of someone who is already very rich, and who may not even spend very much time here (such as sheiks or oligarchs).[15]

Legally, it appears that most of the land, and therefore its habitats, belongs to someone, not to 'all'. If that land is privately owned, its legal owners can (and do) uproot, drain, fell, lease to mining concessions, slash and burn or otherwise destroy with impunity habitats they own. This includes land colonised by rare orchids – and has included the deliberate destruction of orchid colonies because the landowner didn't want orchid geeks coming to see them.[16] In fact, thousands of Britain's designated wildlife sites are on private land without any legal protection.[17]

So, if the country's habitats do not legally 'belong' to many of us, and their owners can treat them as private not collective property, to whom do this land's creatures – the badgers, bats, bacteria, beech trees, birds, Bird's-nests and every other life form which grows, disperses its seed and can fly, lope, wriggle and hop across human property boundaries – really belong?

The CPS definition states they 'belong to all', but if that were so, poaching would not be a crime and the law would not put obstacles in the way of citizens like Sylvia and me from conserving rare creatures in the best way we can. What makes this contradictory state of affairs harder to take is that the laws intended to prevent the habitats and wildlife that (allegedly) belong to 'us' from slipping away are, unlike the trade in illegal narcotics, rarely enforced.

After the police encounter, Sylvia and I drive on in silence until we park and, with more wariness than usual, begin planting out, in potentially self-sustaining mini-colonies of four, the small, slightly carroty rhizomes of Heath Fragrant Orchids, a species which forms steeples of pale violet flowers with a delightful clove-like scent (one of Sylvia's greatest wishes is to bottle that scent as perfume).

She keeps lookout while I use a wallpaper scraper to create a slit large enough to drop the dormant tubers inside, careful to position the nub of next year's growth skywards before letting the opening in the soil close over, and hoping they will survive into next year and the year after that and all the subsequent decades, being pollinated, scattering seeds and, slowly, spreading.

Sylvia has her phone in hand, half watching it, half checking the road. As I guerrilla-plant orchids by the war memorial, she reads to me about the recent approval of a new housing development: 2,500 homes south-west of a nearby city.

'They've allocated a SANGS,' she cynically informs me, with the ironic tone we use for these things. 'SANGS' stands for Suitable Alternative Natural Green Space. I suppose a developer's and the council's definitions of 'natural' differ from ours. It will be 'green', but I am not sure how 'natural' it will be . . . but then, I remind myself, how 'natural' is any of this land after all the tree-felling, quarrying, mining, grazing and many other uses to which it has been put over the centuries? Our forefathers gardened and ploughed, tolerating and encouraging some creatures, driving others to extinction, but now when it comes to new developments, destruction of the natural world seems to be considered as unavoidable. On the whole, people seem too ready to accept that homes built in this way, lives lived in this way, technology used in this way, where the economy works in this way, in a world exploited by people in this way, in which the natural world suffers, is the only way to be.

As such, it is down to Sylvia and me to do what we can at that forthcoming housing development. Sylvia will check the environmental impact statements and vegetation surveys; we will search for orchids in need of salvaging. Later, when the houses are built and the SANGS turfed over, we will return to plant some species there. Now, though, it is autumn, a good time for planting orchids, not seeking them. We must wait until early next summer to save any refugee orchids; in the meantime, we can only hope excavators do not move in. As usual with orchids, it is a race between two different times: orchid time, of seasons and cycles, planet and sun; and human time, a rush to meet deadlines, financial year-ends and stakeholder dividends.

Sylvia's commentary on the proposed site continues as we move to the margins of the rugby pitch.

'The council's trying to unlock millions from the Housing Infrastructure Fund to move the main road, add access to a roundabout, build a bridge, a main road, add four new junctions to the existing road. Blah blah blah.'

It sounds to us like a litany of government-sanctioned destruction. The CPS will not investigate. And this is only one development in one area. Across the country there are hundreds of such developments.[18]

I score shallow trenches near the pitch's dead-ball line; pour seed inside, close the trenches over. Say a silent farewell and wish the embryos good luck. We move on.

With the sound of the Sunday service coming through the village church's thick old walls, I drop a few more dormant orchids next to some granite steps leading into the graveyard. It seems the nation has a clear idea of British history and heritage in terms of bricks and mortar, definable events, cultural acts, monuments and artefacts, and a sense that they belong to us 'all' and must be protected and passed on to future generations. However, although it is woven into this history, the centrality of our living land and the wildlife in it, which the CPS says belongs 'to all', is not afforded the same level of respect or protection. 'Our' land is the backdrop to so many English novels, poems and films; it is seen in traditional embellishments such as Christmas mistletoe, wedding bouquets and funeral wreaths, as well as national symbols such as English oak, Scottish thistle, Irish clover and Welsh daffodil. Culture, heritage, national identity, tradition and the natural world cannot be separated but too often they are. The law does not protect them in the same way. Society rarely views them in the same way. Britain's orchids too are heritage: natural heritage, older than rugby, older than Britain, older than Christianity, older than humanity – but a colony of orchids can be destroyed to make way for car parking while the chances of the same happening to a museum, library, or war memorial are much, much less.

Strangely, on some level, the CPS recognises this connection between cultural and natural heritage. In England and Wales, wildlife crime is bundled up with rural and heritage crime and each CPS area has its own 'Wildlife, Rural and Heritage Crime Coordinator'. Heritage crime *concerns* damage to listed buildings, scheduled monuments, wreck and World Heritage sites, and Conservation Areas. Yet, still the CPS concept of 'heritage' omits heritage in the broader sense of how much Britain's landscape and wildlife provide cultural and financial value to, in the CPS's words, '*this and future generations*'.

Of course, England and Wales are not alone in this. In April 2019, Notre Dame in Paris caught fire. There was international outcry. Countries, institutions and individuals around the world pledged financial aid for its reconstruction. One can imagine the same happening if Westminster Abbey suffered a similar fate. The felling of ancient woodland does not attract the same response.

The falling populations of all Britain's wildlife, due in no small part to the disappearance and degradation of native habitats, are evidence that, despite their ties to our cultural heritage, the *natural* heritage of these lands, the 'creatures and habitats that belong to all' are being allowed to vanish.

If there were any kind of political impetus to push these issues higher up the agenda, if legislation were made fit for purpose, if *individuals* were held to account for the degradation of the land, if there were not a crippling lack of nationwide monitoring, resources, sentencing guidelines for courts and specialist knowledge for police forces, then maybe Sylvia and I would not have to risk prison to try to ensure that one orchid-shaped thread of this nation's rich tapestry, where nature, culture and heritage intertwine, is not allowed to unravel for ever.

14

KENTISH MONKEYS

In the process of carrying out my own guerrilla-planting activities I learned that there is a precedent for this kind of action, a precedent involving one of the rarest orchids in Britain: the Monkey.

Monkeys only became Monkeys at the end of the eighteenth century, when the French botanist Jean-Baptiste Lamarck named the species *Orchis simia* ('Simia' from the Latin for 'of the ape/monkey'). When it first appeared in English in Gerard's *Herball*, it was called the '*Cynosorchis Major Altera*' or 'White Dogstones', its flowers described as 'like an open hood or helmet, having hanging out of everie one as it were the bodie of a little man without a head, with armes stretched out, and thighs straddling abrod, after the same maner almost that the little boies are woont to be pictured hanging out of *Saturnes* mouth'.[1]

At that time these jumbles of small pink-white mottled monkey-men blossomed in good numbers in Kent, Surrey, Berkshire and Oxfordshire. However, soon after receiving their Monkey moniker they were forced to capitulate to increased use of the plough, avaricious gardeners and the grazing habit of growing numbers of rabbits. Monkeys began to flirt with extinction.

Kent's Monkeys were first described in 1777 at a site near Faversham by Edward Jacob, a surgeon, book collector, amateur botanist (and as far as I can make out no relation). In 1802, Monkeys were reported in the same general area but at a different location (they had disappeared from the first). Then, for over a century, the Monkeys vanished. In 1920, they popped up somewhere else in Kent. A few plants flowered at that site for three years, then . . . nothing. The Monkey was presumed extinct. Again. Thirty-two years passed. In late May 1955, Hector Wilks, a chartered surveyor and keen amateur botanist, came across a single plant on a grassy slope. The plant was 10 centimetres tall and had a jumble of thirteen monkey-shaped flowers.[2]

Wilks did not share his discovery widely, nor did he limit his actions to monitoring the plant's fate. He waited for it to produce seed. Unfortunately, while the seed pods were maturing, the plant disappeared – 'eaten', Wilks laconically noted, 'perhaps by a slug'. In 1956, four more plants appeared, but the flowers shrivelled in the heat and no seeds formed. The following year saw the number of Monkeys increase again, but of those on the pasture, two were eaten by horses, one was 'accidentally' picked and the rest, succumbing to another hot, dry season, failed to produce ripe seed. Fortunately, that year dozens more plants were discovered in nearby woodland. Unfortunately, most of them grew in shade, so only three produced flowers. One of these plants rotted; one was mollusc-grazed by a slug;[3] the third produced seed. Wilks collected and dried the seed and, in an effort to conserve and reintroduce this rare species, he scattered it in 'favourable places' nearby. In 1958, a wet spring helped the original population; many more plants flowered. Wilks hand-pollinated them and although one, together with its seed pods, was eaten by a horse, seven produced seed. Again, Wilks gathered, then scattered the seed, some around the original population, some in six other locations.

It's been said that Wilks's initiative wasn't particularly successful.[4] True, of the sites where he tried to introduce Monkeys, they only established in one, but today there is only one other publicly accessible site to see Monkeys in all of the UK. Taking that into consideration, one success is considerably better than zero.

Wilks, as chair of the Kent Trust for Nature Conservation, protested at the irreversible damage to wild orchid sites that the construction of the Channel Tunnel would cause, but his protests were ignored. I can't help thinking that, via the Monkeys, he had some kind of poetic payback. It was more than enough to inspire me to salute his dogged perseverance with a pilgrimage to see the Wilks Monkeys.

It was the first week in June, prime flowering time for Monkeys. I turned north off the M20 and let my phone's satnav guide me through the undulating hills, hedges and villages of a part of England I'd never visited before. Somewhere north was Canterbury; east was Dover. The scrolling map on the little screen took me deep into the Downs and along a narrow country lane. It then informed me it had lost connection.

Lessons learned from visits to other remote orchid colonies had taught me to expect this. Feeling rather smug, I turned to the map and compass I had ready on the passenger seat, just in case. I scanned the map. Checked the compass. Peered outside. The lane seemed to burrow beneath overhanging boughs. There were no landmarks to get my bearings. I couldn't remember the name of the last village I had passed. I didn't know where I was. My smugness sank. I could turn back, but the phone's last suggestion had been to follow the lane, so I figured I might as well do that and, if the phone signal didn't return, hope to find somewhere marked on the map or someone to ask for directions.

The lane wound, dipped and climbed through woodland, then swooped out of the trees to dissect an open patch of pasture. A car was parked at the roadside. A steel gate sealed an opening in the hedge. Next to that was a metal sign. I pulled up. The sign announced I had reached Park Gate Down, also known as 'The Hector Wilks Reserve', a name it had been given by Kent Wildlife Trust in 2007 in honour of the man himself. Smug again, I parked.

Beyond the gate a shallow grassy valley dotted with hawthorn bushes stretched to a gloomy border of pines. I had expected the hidden valley to be surrounded by pristine countryside, but no. Pines aren't native to England. The tall spires that hemmed in the valley gave it the feel of a threatened oasis, not a very big oasis at that.

Inside the reserve half a dozen Greater Butterfly Orchids greeted me. Tall and stately, their pale yellow flowers look nothing like butterflies. Nevertheless, their 'Butterfly' moniker has stuck since 1597 when Gerard described their flowers as resembling 'a white butterflie hir winges spread abroad'.[5] 'White Angels', a name used for them in Somerset, more accurately conveys the appearance of their cruciform flowers,[6] but to me they look like winged serpents taking a leap of faith – if those serpents had been fashioned by Dalí out of lemon cheesecake-coloured wax.

It was late morning. Greater Butterfly Orchids (like the closely related but rarer and just as inaccurately named Lesser Butterfly) are best encountered by moonlight. That's when their flowers become luminous and their lily-like scent strong, all the better to attract their pollinators: large, night-flying moths. Drawn to their ethereal glow, scent and nectar, the moths hover in front of each flower, uncurl their long probosces, slip them into the long spur (the flying snake's body) of each flower, and, to steady themselves, rest their foreheads against the sepals and petals which curve over the opening to the nectary. That's how the flower's pollinia get neatly attached to the moth's eyes or proboscis (where, once attached, they make a double movement, inwards and forwards by 90 degrees).[7]

Next to the Greater Butterflies I found faded flower spikes of Early Purples, the tasselly green flowers of Common Twayblades, the speckled columns of Common Spotteds and flower spikes of a large colony of Chalk Fragrant (still tightly furled in bud). In the middle of the pasture a single, tall Lady Orchid was blossoming like a frilly burgundy-and-ivory beacon, each flower resembling a lady in a red bonnet and a wide skirt.

At the Lady's feet, among the buttercups, docks, red clover and grasses, cavorted a few dozen Monkeys.

When one's only experience of an orchid has been through photographs, encountering it for the first time in real life delivers a shiver of excitement and wonder. The Monkeys were no exception. Each plant had a jumble of white-purple flowers; each flower had a short 'tail', long curly 'arms' and 'legs', a purple-spotted 'torso' and a 'head' capped in a tapering purple 'hood'. Unlike the Greater Butterfly, the Monkey was well named: the resemblance was clear. They were marvellous, complex, cheeky jumbles of candy-floss-coloured anthropomorphic blooms. They were also a link back to Wilks who, to save them, had used methods that, if he had carried them out thirty years later, would have broken the law.

The impressive, varied orchid population in that sheltered valley made me wonder whether it was also a glimpse of an England closer to that experienced by Gerard, when it seemed every pasture, heath, bog and valley had been resplendent with orchids. I reminded myself that the orchids in that site only existed because of unseen, healthy orchid mycorrhizae, and *they* only existed because people who cared about orchids and Wilks's legacy had chosen to protect that valley, now also home to woodpeckers, nightingales and multiple species of moths and butterflies.

15

SAINSBURY'S SLIPPERS

In a village where stone houses cluster around a bend in the Wye, a tall man, youthful-looking (even though he informs me he retired a week earlier), hands me a plain brown envelope. It has the soft, padded texture and weight of an envelope containing another envelope, and, inside that, another, like a not particularly excitingly wrapped game of pass the parcel – unless your passion is saving orchids. If that's the case, then there's quite a prize at the centre. Ash (not his real name) assures me that it contains seed pods of the critically endangered Lady's Slipper Orchid (*Cypripedium calceolus*).[1]

In terms of conservation tales, not just native orchids, the Lady's Slipper is legendary. A stunning plant, albeit smaller in stature and flower size than many of its admiring images make it seem, with a sun-yellow, gorgeous yet faintly obscene plum-shaped pouch, burgundy hood and cascading, twisting sepals like droopy ponytails, it is easily as appealing as any showy tropical species. It was once reasonably common in northern England. However, in the late eighteenth century, when horti-culture became a more popular pastime, covetous gardeners started digging it out of the wild, carrying it home and attempting to grow it. At that time, Smith and Sowerby included it ('the queen of all the European Orchideae') as the first entry in the first volume of their monumental *English Botany*.[2] Smith also noted that, by that time, it was 'confined to some remote and little frequented woods in the North of England'.[3]

Less than a century later, when the third edition of *English Botany* appeared (in 1869), the editors noted that the Lady's Slipper was 'now nearly if not quite extinct'.[4] The wild population had plummeted. It seems most of the home-grown plants removed from their wild habitats had died. The species was declared gone for ever in 1917.

Thirteen years later, two men out for a walk stumbled across a single wild plant on a remote Yorkshire hillside.[5] For decades, secrecy surrounded its location. A group of dedicated guardians was formed (the Cypripedium Committee) and to this day, during flowering season, that lone plant, now well over a century old, remains under constant guard, tripwires alerting wardens of anyone's approach. Even those who know about it are requested to stay away to avoid damaging the surrounding habitat.

For all these reasons I have not handled its seed before. Besides, it is a species which needs cold winters and limestone-based soil, neither of which exist in the right combination in the area where I most frequently operate – along the M5 corridor, as far north as Gloucester with forays into Wiltshire, Dorset, Somerset and Cornwall. In reality, any thought of where I might reintroduce Lady's Slippers should be a long way off. I have to germinate this seed and, where this species is concerned, that is tricky. Nevertheless, I have been tempted here by a curious urge to see if I *can* germinate them, alongside a wistful dream of starting a colony of these exquisite orchids far outside their usual range.

Nothing specific has been stated about where this seed is from. Online, Ash told me it is from an English plant, but, meeting him for the first time, his lively eyes have a trickster glint. It makes me suspicious. I am not sure whether it is *more* suspicious that he has shown up or less. Ash only knows that I am an enthusiastic propagator of native orchids, and he is offering me the seed for free. He offered it to me during one of our more recent exchanges.

Fancy some Cyp. calceolus seed?
Seriously?
Sure.
English?
Yes.
Where? When?

Saying anything online is easy. But here I am, further north than usual and a little wary that I might be walking into a sting; so wary, in fact, that I parked a five-minute walk away so he wouldn't see the number plates; so wary that I drove several hours to collect it in person rather than entrust him with any address I might be connected to. In any case, the postal system's sorting and franking machines easily crush orchid seed. Sending seed through the post has to be done right, preferably in a small glass jar inside a well-protected cardboard box. Even some specialist suppliers don't get it right, meaning the seeds jostle against each other in transit, crushing each other to smithereens, jagged fragments under a microscope, useless.

To my relief, Ash appears to be alone. It occurs to me he could be wearing a wire. Maybe we are being watched. Maybe I am paranoid. Maybe not.

He too is scanning the houses, restlessly shifting his weight as if ready to dash to safety. I wonder if our meeting feels a bit like being in the resistance during a war: a lot unsaid; unspoken fishing for unconfirmed suspicions; the hope that this could be a brother in arms tempered by the fear that it could just as easily be an informant.

'Two green pods,' he says.

He watches me studying the plain packet. Lady's Slipper seed germinates best in a petri dish when it is removed from the seed capsule *before* the capsule matures and dries. No one knows why, because it wouldn't happen that way in the wild. That is what he means by 'green pods' – they haven't dried.

My wariness about Ash leads me to a snap decision: I open the envelope.

'What are you doing?' he hisses, eyes widening.

'Taking a look.'

Ash glances around as if a SWAT team might rappel out of the low, grey clouds. He mutters, drifts slightly closer to shield what is in my hands from the non-existent passers-by. His reaction reassures me.

'Don't drop it.'

I am holding too many envelopes – one from inside the other and another on its way out. He takes two off me. At the core of the manila parcel is a small, rounded, lumpy, coffee-filter-paper package folded in a way which leaves no seams for seed to escape, fastened with a single piece of tape that no escaped seed could stick to. Ash knows what he is doing. This raises questions. Exactly how much does Ash know?

For some time I have been aware that, every year, conservation charities, trusts, societies and teams of volunteers monitor orchid populations. Their work involves managing remnant habitats for the benefit of these increasingly rare plants, putting cages around them to stop hungry rabbits and deer and gaining the necessary official permission to legally rescue and transplant some threatened plants. On occasion I have joined them. Seeing them giving their free time to save orchids has led me to suspect that, somewhere out there, someone else might be doing what Sylvia and I do, whether it is breaking the law, scattering orchid seed or helping wild orchids with a bit of hand-pollination. There are many ways others could be surreptitiously working to save these flowers. Could Ash be a lone wolf? Another orchid outlaw? Or is he part of a wider covert network? These questions dissipate when I unfold Ash's handiwork.

Two large orchid seed pods lie inside. Similar to pods I encounter every summer, but bigger, glossy, green, curved, like very small marrows. They are, I guess, Lady's Slipper seed capsules, but that doesn't mean they are *the* Critically Endangered British Lady's Slipper. They could be capsules from a tropical Lady's Slipper. They could be from a continental Lady's Slipper, illegally imported from some of the few remaining colonies in the Alps. I rewrap them.

'Looks good,' I say, playing it cool. Excited anticipation is building inside me. I want to shake Ash's hand as much as I want to hook him up to a lie detector. I want to get home and start attempting to propagate these seeds. I know caution is wise. I won't know if the seeds are viable until I check them under a microscope, and, as Lady's Slipper seed needs to be sown directly from the immature pod, I won't get to check that until I have the lab prepared, but as my lab is a long way from having the capability to run DNA checks, I simply have to propagate these seeds, then wait, and if all goes well, a decade or more from now, the plants will flower and I will see which species this is.

We wave and nod an awkward hurried farewell.

In the car, pods stashed out of sight, apparently not being tailed or in danger of being descended upon by a specialist Environmental Crime Unit (if only), I head home along the motorway, wondering what I am going to do with those seeds, and how Ash got hold of them.

Although Ash and I, and maybe others, are Wilks's ideological descendants, clandestine renegade protectors of some of Britain's most threatened wild flowers, the future of the world's orchids does not rest entirely in our hands, or in those of the volunteer teams who go to their rescue every year. There are some big players in the orchid-conservation world, probably one of the most renowned being the Sainsbury Orchid Conservation Project (SOCP) run through the Royal Botanic Gardens at Kew. After collecting the alleged Lady's Slipper seed from Ash, I began to suspect this project might explain how he had legally obtained British Lady's Slipper seed pods, but I wanted to be sure.

The SOCP had inspired my dealings with orchid seed, but my knowledge of what actually went on – and goes on – in the project at Kew is sketchy. Secrecy has surrounded orchids ever since the days when Victorian orchid hunters vied to discover rare tropical species. In recent years, differing ideologies have clashed: education versus discretion; the long-term benefits of sharing the locations of protected species and teaching visitors about the importance of their conservation versus maintaining complete secrecy to avoid the unwelcome attention of plant thieves, vandals and visitor pressure on fragile habitats.[6] Details of what goes on at the SOCP are scarce (not least, I assume, because of the importance of protecting research findings prior to publication). However, to try and understand the provenance of the seed Ash handed me, I did some digging. Sylvia and I found some names, made inquiries and assured those we contacted that they needn't go 'on the record'. In most cases, we were met with, at best, cryptic references to fungi and more questions ('What do *you* know about what goes on there?'). More often, the response was no response at all. Secrecy seemed to still be very much at play. Why?

The financial backing for the project began with Lady Lisa Sainsbury, wife of Sir Robert Sainsbury (whose grandfather founded the Sainsbury's supermarket empire), and her sizeable tropical orchid collection. In the late 1970s rumours circulated that this collection was going to be dispersed. The Royal Botanic Gardens approached Lady Sainsbury and suggested she might consider Kew as her orchids' new home. In 1980 Kew received her collection.[7] In 1983 the Sainsburys funded an Australian botanist, Mark Clements, to spend eighteen months at Kew researching methods of growing British orchids from seed.[8] Two years later, the Sainsburys donated more money, this time to fund the Sainsbury Orchid Fellow, a research post, and the Sainsbury Orchid Conservation Project. The project's objective was to invite 'research into the cultivation of endangered species, with emphasis on British and European orchids . . . to re-establish them in managed sites in the wild'.[9]

By 1987 Kew established its first trial planting of Kew-grown orchids in an area where orchids already grew in the grounds at Wakehurst Place (now known simply as Wakehurst), a manor house in Sussex also home to Kew's Millennium Seed Bank. The introduced orchid species was the Loose-flowered or Jersey Orchid (*Anacamptis laxiflora*).[10] In Britain this species is only found on the island of Jersey, so it might seem an unusual choice. The decision was deliberate: Jerseys would be easily distinguishable from the other native orchids already growing at that site.[11] By the end of 1989, 350 Jerseys had been introduced; eighty flowered.[12] In the same year the Sainsburys donated an additional £1 million,[13] and four other Kew-propagated native species – Common Spotted, Southern Marsh, Bee and Green-winged – were introduced to the grounds of Wakehurst and Kew;[14] a few were donated to a flowerbed outside the Nature Conservancy Council's head office in Peterborough.[15]

I wondered why none of these were particularly rare species. Apparently the intention was to monitor their progress in order to learn how best, in due course, to introduce rarer orchids elsewhere.[16] As a scientific approach, this seemed a little rudimentary. If you want to know how a Fen Orchid, say, might best be propagated and reintroduced, you collect data from monitoring Fen Orchids, not Bees, Jerseys, Common Spotted or any other species. Fen Orchids grow in marsh, do not have tubers and can be pollinated by rain;[17] Bees (and others) do not share these characteristics. Likewise, Jerseys are very limited in providing useful information about Lady's Slippers; Green-winged, Bog Orchids and so on.

Kew had managed to propagate these species by isolating a fungus (named only F414) from the European Crimean Marsh Orchid (*Dactylorhiza iberica*), but this fungus could not germinate all orchid species.[18] Not for want of trying, Kew scientists had failed to isolate the mycorrhizae needed to germinate some of Britain's rarest orchids, including the Lady's Slipper. To remedy this, Margaret Ramsay, head of the SOCP at that time, travelled to Sweden to learn how Svante Malmgren had succeeded where

Kew could not. Ramsay found an amateur enthusiast who, in five years, had produced hundreds of Lady's Slipper Orchids in his kitchen without even an air-purifier hooked up to a plastic box (Malmgren carried out his micropropagation over a pan of boiling water and stored the young embryos in flasks in a box in his bedroom wardrobe).[19]

Ramsay learned as much as she could from Malmgren, returned to Kew and adopted his germination technique. By 1992 Kew announced that five orchid species (almost certainly the aforementioned Jersey, Bee, Green-winged, Common Spotted and Southern Marsh) had been 're-established in wild situations', and another forty-six British species had been germinated in the laboratory.[20] That would equate to basically every recognised native British orchid, an extraordinary achievement given that certain species (in particular the helleborines) are notoriously difficult to propagate artificially. How Kew managed to do that has not (to date) been publicly shared.

The Harraps report that, four years later, Kew introduced ten rare Monkeys to a site in Oxfordshire to boost the population there.[21] In 1997, the SOCP was working with other rare species, including Military and Burnt,[22] and, in the same year, seventy-five Kew-grown Fen Orchid seedlings were reintroduced to the Norfolk Broads.[23] These reintroductions were not a runaway success. By 1998, of the ten Monkeys, there was sign of only one. Of the Fens a study published in 1998 does not mention reintroductions; similarly, in 2018, when Jon Dunn wrote about the Fen Orchids at the same site, he noted only that Kew and other conservation bodies were closely monitoring their steady decline (not that any plants had ever been reintroduced).[24] It appeared that all that funding and expertise had not translated into orchid-reintroduction gold.

But what of the Lady's Slipper, the flagship reintroduction species, the reason why Ramsay travelled to Sweden, the seeds of which I may have obtained from Ash? It emerged that, although there was only one surviving, publicly recognised, seemingly 'wild' Lady's Slipper in Britain, at least two other Slippers had been found in gardens in northern England: one in the market town of Kendal, in Cumbria; the other in (or near) Hornby, a village in North Yorkshire. Imaginatively referred to as the 'Kendal' and 'Hornby' plants, they were thought to have been transported from the wild into gardens in the nineteenth century.[25]

It was fortunate that they existed for, as Darwin's orchid experiments had shown, self-fertilisation in orchids can result in less viable seed and weaker plants. From the late 1980s through the 1990s, working in association with the SOCP, Britain's wild Lady's Slipper was painstakingly hand-pollinated and cross-pollinated with the two 'domesticated' plants. Pollination was successful. Green seed pods were picked and taken to Kew's laboratories, where Malmgren's asymbiotic propagation technique worked like a charm. The hundreds of plants produced were destined to be part of English Nature's Species Recovery Programme and would be reintroduced to sites across northern England, relieving pressure on the wild plant, reversing centuries of the species' decline and bringing it back from the brink of extinction.[26] At least, that was the plan.

Mortality rates were high among the first reintroductions – 'mollusc grazing' a particular hazard – but Kew's laboratories continued churning out seedlings, so many that by 1998 they sought amateur orchid enthusiasts to take in their surplus Lady's Slipper seedlings, asking them to note down their culture methods, their level of success, and allow that data to be collated to assist the reintroduction project.[27] Forty-two members of the Hardy Orchid Society received Kew-grown seedlings.[28]

There, I thought, lay the answer to how Ash got that seed pod. The plant would be twenty-plus years old, a relative youngster for a species that has been recorded as living for nearly two hundred years.[29]

The first flowering of a Kew-grown Lady's Slipper in the wild occurred in the early years of the new millennium.[30] More flowers followed. Success appeared to be dawning. By 2014, reintroduced Lady's Slipper Orchids had flowered at eleven sites, but . . . eight years on, many of the Lady's Slipper colonies had become neglected and overgrown. Others had simply become too dry. Climate change was a contributing factor, so too an error in judgement by whoever chose the reintroduction sites: they were looking for locations that closely matched the habitat of the single wild plant, but, with time, it was gradually accepted that that plant could be anomalous; Lady's Slippers do not usually grow in that kind of relatively dry habitat.

Gait Barrows, a nature reserve in north-west England, bucked this trend. Thirty reintroduced Kew-grown Lady's Slippers were successfully established there; many flowered every June, but the last time any flowered was in 2021. In January 2022 the senior manager of Gait Barrows announced online that every Lady's Slipper in the reserve had been removed, along with those at seventeen other sites.[31]

I was incredulous. Had they been victims of mass theft? Was this intentional disinformation designed to reduce the number of visitors to the site? It turned out to be no joke: the plants had been removed the previous autumn.

Genetic data had revealed that, although the same species, the Kendal plant used to cross-pollinate the wild British plant was 'of continental origin'. It was assumed that, probably in the nineteenth or early twentieth century, that plant had been transported to England from Europe. Concerns were raised around how further cross-pollination might reduce the native genetic provenance of the wild English plant and produce plants less suited to the British climate. After decades of research, work and a big wad of cash, the established Lady's Slippers at Gait Barrows – and those seventeen other sites – were dug up; those at Gait Barrows were given to volunteers to care for 'in captivity'.[32]

★

This news left me perplexed. Natural England's (and Kew's?) views about genetic plant purity were understandable on an aspirational scientific plane where everything fits into neatly defined boxes, but the reality orchids had shown me was that the living world more usually merges, connects and interdepends. Using genetics, as Kew was suggesting, as justification for digging out those plants had – to me – a whiff of botanical eugenics. What's more, the justifications given for the orchids' removal seemed illogical. The Gait Barrows plants had been established successfully, suggesting they *were* reasonably well suited to the habitat.

What's more, genetic resilience is being challenged in so many British plants by a world undergoing rapid climate change that an argument could be made that continental genes might *increase* resilience to an unpredictable climate. Might it be argued that the two genetically different species had freely interbred for decades, with bees moving from imported plants to wild plants without a scientist asserting that what they were doing was not allowed? In this regard, it seemed that those rare orchids were being subjected to a level of genetic screening inconsistent with other reintroduction programmes. 'Continental origin' had not been an issue when Scotland reintroduced beavers from Norway, southern England reintroduced Great Bustards from Russia and Spain, or even with Kew's introduced Jerseys at Wakehurst, because *that* seed had come from Crete.[33]

I emailed Kew to see if they could explain more about their rationale for pulling the plug on those Lady's Slippers, particularly in light of conservation projects with other species where continental genes were not considered problematic. A helpful member of Kew's PR crew responded, saying a professor, an expert in orchid genetics no less, who had worked with the SOCP since the 1980s, would be in touch. Eagerly, I awaited the professor's email. It didn't arrive. A month later I tried again. Nothing. I found the professor's Kew email address and politely sent my query direct. Aside from my question about Lady's

Slipper genetics, I added an enquiry about how many British native orchids the SOCP was actively involved in reintroducing, because I hadn't found much news about that since the early 2000s. I waited. No response. I tried again, sending from my official-looking day-job email address. I never received a reply. I gave up.

Either Kew had very rigorous email filters or the professor in orchid genetics had far more important things to do than answering random queries from a stranger. I wasn't going to harass him and I don't hold his silence against him; it simply means it feels presumptuous to offer conclusions of my own without being able to take into account more of a response from Kew. Nevertheless, based on the facts I have found concerning more than thirty-five years of work, in terms of *reintroductions*, it seems that the Sainsbury Orchid Conservation Project has had, at best, very modest success.

Perhaps Kew has undertaken thousands of successful (but very clandestine) orchid reintroductions and they just don't want to talk about it. A 2022 study which searched decades of sightings by members of the Botanical Society of Britain and Ireland suggests that isn't the case: it shows that populations of all British orchids are in steep decline, with recorded sightings of orchids in Britain falling by an average of 60 per cent since 1950 – and this despite more and more sightings being provided to the society.[34]

Without taking away from the importance of Kew's wider research and conservation projects, the SOCP's lack of success in what it set out to do *might* explain why it's difficult to find information about what is going on there now, and why those who have been involved prefer not to answer awkward questions from an amateur enthusiast. After all, I imagine that Kew quietly uprooting its flagship Lady's Slippers because they had the 'wrong' genes is hardly the kind of PR coup the project needs.

That said, although data is sparse, SOCP reintroductions *do* continue. Hundreds of Kew-grown Green-winged Orchid seedlings (not one of the country's rarest) were planted at Gait Barrows in September 2021 and there is hope that 'true' British Lady's Slippers will be reintroduced there once again – no one knows when. Undoubtedly, Kew's research has positively influenced orchid conservation around the world and raised awareness of their plight, but, several decades on, the Sainsbury Orchid Conservation Project's triumphant success has yet to materialise.

16

THE SAWFLY ENIGMA

A scene in Ian Fleming's novel *On Her Majesty's Secret Service* sees James Bond enter his boss's study and notice 'an extremely dim little flower' in a glass of water. Reluctantly setting aside his unfinished watercolour of the flower, Bond's boss, M, explains that the little flower is an orchid, but not one of those 'disgusting' tropical species, which are 'damned near animals, and their colours, all those pinks and mauves and blotchy yellow tongues, are positively hideous'. M explains that the orchid in the glass is 'the real thing', an Autumn Lady's-tresses obtained especially from 'a man I know – assistant to a chap called Summerhayes who's the orchid king at Kew'. M elaborates on the flowering times of Autumn Lady's-tresses and the research carried out by the 'assistant' on orchid mycorrhizae.[1]

Bond's creator had a reputation for gambling, heavy drinking and chain-smoking. It may seem uncharacteristic that he knew anything about Autumn Lady's-tresses, let alone bothered to include them in Bond's action-packed adventures. Yet a side of Fleming was attuned to the natural world: he walked in woods, collected wild flowers and borrowed the name 'James Bond' from the author of a book he owned, *Birds of the West Indies*. So a meeting between Fleming and the real-life 'orchid king at Kew' while he was writing this Bond novel – that 'orchid king' really was called Summerhayes, Victor Summerhayes – would not have been entirely surprising, and nor is its inclusion in Bond's world.

M's hobby of painting watercolours of Britain's wild orchids is a useful device to illustrate M's patriotism and it establishes that M, like Bond (in a more understated way, rather like the orchid he is painting), has an anti-establishment streak: he elevates demure, underappreciated British orchids over 'hideous' establishment favourites. M is presented as a champion of British orchids, albeit a fictional champion. However, Fleming's reference to the *real* Summerhayes, an expert in British orchids and head of Kew's orchid herbarium between 1924 and 1964, lifts his orchid reference out of the pages of his novel and into 1960s Britain, where a loose network of sometimes mildly eccentric writers and botanists were operating, all of them passionate about 'the real thing'.

Aside from Victor Summerhayes, whose book *Wild Orchids of Britain* (1951) was a standard reference on the topic for years, another member of the network was the writer Jocelyn Brooke. A school truant who was also expelled from Oxford University, Brooke served in the Second World War as a 'pox wallah', an army medic tasked with treating venereal disease. From the age of four, Brooke had been keenly interested in Britain's orchids. The novels in his semi-autobiographical *Orchid Trilogy* (1948–50) interweave accounts of 'his' search for orchids with other events in 'his' life.

In 1950 he published a non-fiction book on Britain's orchids in which he described the network of orchid lovers operating in England at that time as 'few but fanatical', a kind of 'cult . . . a freemasonry with all the jealously guarded secrets of such institutions'.[2] He might just as well have been describing Bond's network of spies. A note in Brooke's *The Wild Orchids of Britain* provides an example of one such secret.

Brooke spent years searching for the Military Orchid, a species thought to be extinct in Britain. He was unsuccessful, but another member of that 'cult', Job 'Ted' Lousley, accidentally discovered it in 1947. Lousley, a banker, had a sideline in botany and amassed the largest private herbarium in Britain (25,000 specimens collected from every area of the country). He had a special interest in studying how plants colonised bomb sites in London after the Second World War. The common dock and invasive species were Lousley's specialities.[3] Perhaps this made his discovery of the Military all the more infuriating for Brooke, especially as, after he discovered it, Lousley refused to tell most people where it was.[4]

In response, Brooke, his frustration evident, wrote in a foot-note, '[The Military Orchid] was rediscovered in Buckinghamshire by a well-known botanist who, embittered no doubt by former iniquities on the part of his colleagues, refuses to divulge its exact whereabouts.'[5]

It took nine years for another 'cult' member, Francis Rose, to track down that Military colony. When he did, he sent Lousley a coded message on a postcard. The message read: 'The Soldiers are at home in their field' – an oblique reference to the Military Orchids' secret base in a Buckinghamshire wood.[6]

In contrast to his opinion of Lousley, Brooke expresses his 'special debt of gratitude' to Rose,[7] a man who was generally regarded as the best field botanist of his, or perhaps any, time. Fascinated by plants as a boy, Rose taught himself biology (it was not offered as a subject at his Catholic school) in order to enter Chelsea Polytechnic to study Natural Sciences. During the Second World War he worked on explosives at Woolwich Arsenal, then returned to research and academia, writing dozens of articles and reports for the Nature Conservancy Council and spending twenty years composing the ground-breaking *Wild Flower Key* (1981). With inexhaustible energy and infectious enthusiasm, he was known for dragging students on hours-long treks without lunch or dinner, but always an eye on pub opening times.

In his turn, Lousley disapproved of Rose, in particular Rose's habit of carrying out 'experiments' involving scattering wild orchid seed around southern England. These experiments caused head-scratching among botanists when orchids appeared outside their 'normal' range. It is 'most unfortunate', wrote Lousley, 'that in 1942 Dr F. Rose sowed seed [of the Lady Orchid] from Kent near the main road up Titley Hill [in Surrey] and failed to keep his experiment under close observation'.[8]

Lady Orchids were not the only species Rose scattered: Lizards, Early Spiders and even non-native Mediterranean species got his 'experimental' treatment.[9] Yet, while Lousley disapproved, Rose's seed-scattering spurred another member of the orchid network, Rose's friend Hector Wilks, to successfully do the same with Monkey seed. Rose may also have been responsible for one of England's recent orchid enigmas.

The Sawfly (*Ophrys tenthredinifera*), gaudy cousin to Britain's Bee, is a stumpy Mediterranean species. Neon green and candy pink, it resembles a new variety of Bassett's Liquorice Allsorts. It shouldn't grow in England, but in 2014 a Sawfly appeared in the short turf above Dorset's coastal cliffs.

In the same way unexpected migrant birds sometimes alight on Britain's shores, had a stray seed blown there and germinated? Had a seed hitched a lift in the treads of a holidaying naturalist's boots? Or had seed scattered by Rose – or other person(s) unknown with flagrant disregard for the accepted code of practice around planting species in the wild – quietly grown, spread unnoticed and produced this singular plant? Or could the Sawfly have resulted from the actions of another orchidophile, the internationally acclaimed writer John Fowles?

Fowles lived in Lyme Regis, a small town close to where the Sawfly appeared. Despite being an 'unauthorised person' in the eyes of today's environmental law, he was known to have rescued Bees from a nearby landslip. He also illegally smuggled boxes of orchids from the continent to his sheltered garden, where they successfully grew.[10] Fowles had a special passion for insect mimics, such as Bees and similar Mediterranean species. Of these, the Sawfly had a particular significance for Fowles, as he revealed during an interview in which he describes lying in a hospital bed shortly after suffering a stroke, filled with self-rage and pity, repeating to himself the word:

> *tenthredinifera, tenthredinifera, tenthredinifera* . . . That unpronounceable name belongs to one of the most beautiful *Ophrys*, or bee orchids, of Europe. I had come upon it on a Cretan mountain the previous spring, and I was saying that name like a mantra because I thought I should never climb that remote mountain again.[11]

In that moment the Sawfly embodied what Fowles risked losing for ever: the seasons, his orchids, beauty, fitness, health, life. Repeating its name was an act of hope, a way of clinging to everything that plant represented.

For Fowles, as for Rose, Brooke, Wilks, Summerhayes and Lousley, despite their disagreements, orchids were more than garden ornaments: they were their passion. Fleming's M was drawn to Britain's orchids because they were the unpretentious 'real thing'; Summerhayes (the real Summerhayes, who fought in the Battle of the Somme, had a penchant for wearing a 'disgusting' hat while botanising and, for reasons best known to himself, chose to pronounce *species* as *svecies*) thought their appeal derived from a combination of fragrance, 'delicacy of colouring', 'fantastic resemblances to other living beings' and, most of all, their rarity.[12] Brooke attributed their attraction to some 'indefinable quality . . . something rather perverse and ambiguous, something even a trifle sinister'.[13]

Sometimes I too wonder what draws me to these flowers. The attraction of tropical orchids is easier to explain: they have a striking, sometimes otherworldly beauty, with connotations of an alluring elsewhere of mountain tops wreathed in equatorial cloud and jungles teeming with life. The beauty of Britain's orchids is often on a smaller – sometimes tiny – scale, perhaps all the more exquisite for being minute masterpieces.

For me, the fact that these orchids are not sold in every supermarket adds to their appeal. Their beauty requires effort to locate, and finding them in Britain's dwindling meadows, woods, heaths and bogs is an increasing challenge. Searching for them has taken me to areas of the country I would otherwise never have seen. Encountering them always brings me joy, just as each time I successfully rescue, grow and guerrilla-reintroduce one, I get a far greater sense of satisfaction and purpose than anything offered by hours in an office filtering spreadsheets. Britain's orchids are quirky, surprising, unconventional. I cherish that and I am grateful to them for opening my eyes to so much. In exchange, I feel a duty to protect them.

Inevitably some of the same elements that attracted me to Britain's orchids were part of the 'indefinable quality' that attracted Brooke, Wilks, Rose and Co. I also wonder whether for those men, survivors of world wars, British orchids offered an opportunity for a kind of gallant quest to protect innocent British beauty from overwhelming odds.

Summerhayes lamented the destructive consequences of 'intensive agricultural use of our countryside . . . the felling of woods, ploughing and drainage'.[14] Most orchids, wrote Brooke, 'are doubtless doomed to ultimate extinction, along with the woods and downlands which they adorn'.[15] Just as Fleming's Bond defended Britain from fictional evils, Rose, Wilks, Summerhayes, Lousley, Fowles and Brooke recognised that Britain's 'real' orchids were facing an existential threat. They did what they could to save them.

Wilks was effective in saving Monkeys. Rose inspired dozens of students, and his *Wild Flower Key* remains a landmark work. Kew now has a Francis Rose Reserve devoted to one of his other passions: lichens. Lousley's huge personal herbarium and notebooks are housed in national institutions, and the Military colony he found continues to thrive. Nevertheless, a human generation later, orchid numbers in Britain and the habitats they need are a fraction of what they were in the 1950s.

As the decades have passed, the threats wild orchids face have increased; unpredictable rainfall and record-breaking heat are now among them. Depending on how one looks at it, this need not indicate a looming Brooke-envisaged doom. Orchids that are adapted to cool, damp conditions and less resistant to drought are suffering, but the Sawfly could be a sign of what is to come. In 2019 that lone visitor disappeared as enigmatically as it arrived, but it is not the only Mediterranean species to have appeared in England in recent years.

In 1989, a small colony of Small-flowered Tongue Orchids (*Serapias parviflora*), more usually residents of the Mediterranean basin, raised flower spikes in Cornwall. The name *Serapias* comes from Serapis, a Greco-Egyptian god who attracted an orgiastic cult of followers. Given the flower's labellum is a single long russet-green 'tongue' which spills from a tight orifice of petals and sepals, this sexual association is probably warranted.

Despite enthusiastic photographers occasionally trampling flower spikes, the exotic visitors flowered and spread. When the grazing regime at the site changed in 2009, the orchids were flattened by cattle, invasive plants moved in and the colony disappeared. It was thought that marked the end of a short-lived incursion by the Small-flowered Tongue. Then, in 2021, a colony mysteriously appeared on an eleventh-floor roof garden in Central London. Meanwhile, the Small-flowered Tongue Orchid's bigger sibling, also a Mediterranean resident, the Greater Tongue (*Serapias lingua*), has been found in South Devon. It persisted for five years spanning the millennium, then disappeared. Another colony, in Essex, seems to have become naturalised.[16] As long as a recent planning application, which threatens to destroy the site and the orchids growing there, is not approved, they may well spread, joining other windblown pioneers more adapted to a warming climate, just as some of Britain's resident species are driven northwards. At some point, of course, if their habitable range continues to diminish, there will be nowhere left on these shores for cooler-loving species to go.

STRIMMERS AND STRATEGIES

The truck exhales a wave of engine snarl and trailer-clatter as it rushes from darkness into darkness at about sixty. High in the cab the driver stares. Did he see me? If so, his foot didn't stray off the accelerator. The truck's *gnarr* fades. Its rear lights recede. It rounds a distant bend. Whispering darkness returns.

I clamber quickly down the mossy slope bordering the road. My right hand clasps a wallpaper scraper, my left a small torch. Trying not to slip, I follow its beam across the spongy ground until I reach a patch suitable for orchids. I lay the torch on the ground and stab the scraper's blade into the land's living skin. With well-practised moves, I lever left and right, widening the incision. From a pocket I retrieve a freezer bag. Inside are a dozen orchid rhizomes. They are Common Spotted, with tapering limbs and little nub heads, a mixture of flowering-sized plants salvaged from building sites and my home-micropropagated additions. The home-grown orchids are two years old, smaller than their wild brethren but old enough to be deployed, and, most of all, they are 'mine'.

Rescuing orchids from a site where they would otherwise be destroyed is one thing; raising them from seed brings a whole new level of satisfaction. Among the first generations of orchids from my kitchen, my fridge, my yard, my effort, Common Spotted are very far from the rarest orchid in the land, but they represent an important step, an act of hope. They prove what I want to do can be done, and there, beyond the city limits, on that mossy bank near a lay-by where a woodcarver occasionally sets up a stall selling giant wooden carvings of eagles, there, I thought, those orchids might escape the council mowing regimes and the herbicide-wielding groundsmen that so often devastate my urban guerrilla-rewilding attempts; *they* had driven me to that spongy bank that dawn.

In the early years, I kept no record of where I planted or which species I introduced; I focused only on trespassing on construction sites at the right time of year (without getting caught), saving orchids, then getting those I didn't want to keep as orchid orchards out of the yard (at the right time of year) before anyone grew suspicious, and planting them (without getting caught).

In predawn light, in drizzle, rain, wind and stillness, sometimes accompanied by the song of blackbirds, I wandered the city, certain by then that most of its soil was a neutral pH, and plunged my trowel into the earthy margins of a cricket field, a patch outside a squash court, a grassy mound beside a block of flats, margins of car parks, verges beside main roads, on round-abouts, along footpaths and in playgrounds. Into the cuts that I made I dropped salvaged orchids.

I seldom returned to monitor my work. If I did, I found occasional triumph and frequent disappointment. Sometimes areas that had looked so promising when I planted them in autumn were swamped by tangled vegetation in May. Any orchids trying to survive there would be starved of light, and their flowers (if they bloomed) would be so hidden in the thicket that no pollinators would find them. In those conditions, most orchids die. Without enough solar energy to store, their new tubers would be very small, perhaps too small to allow the subsequent year's orchid shoot to reach sunlight. Sometimes I found everything that had been growing there – grasses, buttercups, thistles, daisies, clover and the orchids I had planted – mown down and left to rot on the stubble of shorn grass. This was even the case on protected land.

A colony of Autumn Lady's-tresses grew on the grounds of one of the many universities I have visited. The university knew they were there and, around flowering time, erected cordons and signs warning 'Keep Off! Orchids Flowering!' In some ways, that was great – the colony was protected. It also meant I knew that the earth and the fungi in it were suitable for Autumn Lady's-tresses, so any Autumn Lady's-tresses I found could be relocated there . . . which I unofficially did, together with several other species. I thought the university would be pleasantly surprised when Common Twayblade, Greater Butterfly, Bee and Common Spotted started appearing there. So, for a few years I returned to that verge both to plant and to admire the emergence of those delicate spires. Over time, I noticed that the orchid numbers, including those I had planted, were decreasing. The colonies were dying. The reason? The grass there was mown a few times a year, but rather than being gathered up, the grass clippings were left to rot into the ground. That changed the chemistry of the earth. That didn't suit the orchids or the orchid mycorrhizae.

By that time I had joined the Hardy Orchid Society, a dedicated group of volunteers, enthusiasts and experts in the terrestrial orchid world. My experiences to that point had not filled me with confidence that any kind of institution would take the slightest notice of me, so rather than waste time on unanswered emails, I told the society's conservation officer, Bill Temple, about the looming demise of that colony. Bill contacted the university and explained how and when that area needed to be mown (essentially at certain times of year with the grass clippings gathered and removed) to save the orchids. I got copied into the emails. The head of grounds maintenance responded, saying they were already doing all those things. As far as I could see, no cut grass on that area was ever removed. They continued to erect the signs and cordons, but within a couple of years of not following Bill's advice, the signs were warning passers-by of orchids that were no longer there. A year later they stopped putting up the signs. A flourishing colony of rare flowers augmented by several of my additions had died because someone didn't understand the devastating impact of not raking up a bit of grass.

In response to those early disappointments, I devised a planting strategy which I thought would outsmart spinning blades. I planted rescued orchids close to boulders, next to walls, under signs, at the base of lamp posts, on steep slopes, under benches – places where ride-on mowers could not reach and where a lazy strimmer operator might not be bothered to strim.

I started to see success, most noticeably in a park where I planted Common Spotteds around a fallen tree which had become a de facto bench, popular in the summer evenings with teenagers who went there to smoke illicit substances. Sheltered from strimmers and mowers, unexpectedly hot summers and unexpectedly wet winters, the orchids flourished. To this day they flower and seed, and over the years subsequent generations have slowly emerged to take the place of the original refugees. I'm not sure how much notice the teenagers high on cannabis give them, but occasionally I check in on the colony's gradual spread and it reminds me that, despite the setbacks, guerrilla rewilding can be done.

That strategy did not guarantee success, however. Another location was a short detour from my usual walking commute to work, where train tracks running under the city emerged above ground at an unmanned platform. A flight of concrete steps descended to the platform. Along the left side of that descent rose a grey stone wall. Above that extended a steep bank clad in buttercups, ivy, moss and crane's-bill. It seemed the perfect place for a colony of orchids; there, for a few weeks of every year, they could offer a display of intriguing blooms to visitors and commuters. So convinced was I that I returned there several times to climb the wall, balance precariously on top and reach as high up the bank as I could to slip a good percentage of the early refugees a couple of centimetres into the earth.

In the first flowering season success was limited: maybe half the plants emerged and half of them flowered. I considered those few resolute flower spikes a victory. Buoyed, I kept returning there to plant. Although each year some were whittled down by hungry slugs and snails, in May and June of subsequent seasons patches of the slope displayed lilac spires of Common Spotteds and the Twayblade's big twin oval leaves and lime-green starry steeples. A gallery of orchids was forming. I hoped someone noticed.

Then, in the third or fourth year, I returned to the slope in July to check whether any of the flowers had produced seeds. There weren't any. Everything growing there was wilted, yellowed, blighted. It looked simultaneously burnt and drowned. Herbicide. I'd seen a few men in white overalls with plastic tanks on their backs patrolling roadsides as they waved long wands of liquid death over the life unfurling from cracks between paving stones. I never considered they might spray that bank. Very little grew there after that.

Looking back, it might seem mystifying that I didn't contact the local council and say, 'Hey, I'd like to plant native orchids around the city. Will you let me? You tell me where, then you agree to not strim or spray those places.' It seems simple. Reality, however, is less so. In its attitude to what was at stake (to their mind probably just a few common wild flowers) government, wealthy development companies, environmental and planning laws, and law enforcement didn't seem eager to engage with the loss that was going on. In instances where the authority did care – botanical societies, Kew – they didn't seem eager to engage with me. From their perspective, of course, there was the question of legitimacy. I had no track record. If I approached the local council, the same factors would come into play.

I was no one and I would be asking a council to change its working practice. Furthermore, I saw little to suggest the council cared much for plants. Memorial gardens dedicated to some famous local Victorian plant hunters had been allowed to descend into unattractive sprawling masses of bindweed. Litter gathered there. Graffiti obscured the information plaques. Around the city centre, flower beds and hanging baskets were planted; the plants died, were replaced with new ones . . . until they died and the cycle continued.

Stories countrywide told of councils 'accidentally' mowing down habitats set aside for butterflies, orchids and other wildlife, and, of course, councils were selling off land and approving developments despite the huge ecological cost to the nation. Put simply, while there probably were eco-conscious planning officers and council workers, I doubted they were among the high-vis wearing, poison-pumping, mower-riding squaddies tasked with maintaining the city's green spaces.

It seemed simpler just to get on with it, plant by precious plant, with someone (me) I knew was going to get the job done.

Urban guerrilla orchid planting was a thrill and a challenge, but too many setbacks caused by mowers, strimmers, poison wands and carelessness led me to the roadside lay-by. I suspected that intercity routes, locations motorists paused at or passed, still offered good opportunities for engineering displays of orchids while also being places where strimmers and sprayers seldom ventured.

I had passed that mossy bank a few times and it always struck me as a good place for orchids. The thing was, there would need to be a lot of them to make an impact on speeding motorists. Fortunately, since I had started my orchid production line, I had plenty. Time and again I drove out to that slope at night and planted and planted, each time cautiously listening for the approach of a stray vehicle and scurrying back to a hiding place before headlamps caught me. I had good reason to be wary. I could imagine the phone call.

'Hi, Police? I've just passed a lay-by on the A396. There was a guy with a torch and what looked like a big knife. He was acting like he didn't want to be seen. I'm pretty sure he was *burying* something beside the road . . . Did I see what it was? No. It wasn't *big*, like a body, but it could have been body *parts*, or, maybe he was hiding evidence . . . Maybe he was burying the knife . . .'

I could end up being the subject of a manhunt. They wouldn't find a murderer, a weapon or body parts, but they might find illicit orchids. I doubted anything would come of that, but the whole scenario was definitely best avoided, so as usual I kept my orchid planting a secret.

Now, in a good year, two dozen orchids decorate that bank, enough for observant motorists to see them. And I continue planting there. I want the whole bank to be resplendent. I want cars to stop, people to get out, crowds to be drawn there to admire nature's stunning, beautiful complexity.[1]

I have also popped young plants into suitable patches on industrial estates (people care less about strimming there), by park-and-rides, outside farm shops, on hedge-banks beside rural lanes, along water courses, under bridges. For years I have planted with strategic abandon, and as time goes on, when I scan the BSBI database for orchid sightings, I notice that orchids I planted are appearing there. At least someone, even if someone already interested in flora, is noticing. I only hope – perhaps even assume – that people who *aren't* amateur botanists might experience a flash of interest when – if – they notice them.

18

SAMPHIRE HOE

Given that much of my early summer is spent rescuing orchids from destruction while I play hide-and-seek with construction-site security guards, it's ironic that, after the building phase is complete, development sites can offer great opportunities for orchids. This is because some native orchid species are naturally a bit like weeds in that they can be among the first plant colonisers of heaps of barren earth dumped after the creation of underpasses, foundations and tunnels. An extreme example of this can be found where 5 million cubic metres of chalky earth produced by the construction of the Channel Tunnel were dumped near Dover.

Among orchid aficionados, this site is renowned for its Early Spider Orchids. The Early Spider, like the Bee Orchid and Fly, is an insect mimic. Its fuzzy brown labellum, marked with a pale H (which, to my mind, often looks like a cartoonish pair of vampire fangs), has a texture, smell and colour designed to attract solitary male bees.[1] The Early Spider usually produces fewer flowers on a shorter spike than the Bee, but flower numbers and height of spike do not matter to orchid lovers who wait impatiently for the opening of its five-pointed coronas of slim, fresh, lime-green petals and sepals (the spider's legs). That is because, alongside Early Purples, Early Spiders herald the start of orchid season.

First recorded in England in 1650 in Northamptonshire, the Early Spider has vanished from over 70 per cent of its former range, including everywhere in northern England. Giving it Schedule 8 protection failed to save it from extinction in Wiltshire (in 1989), the Isle of Wight (1992) and Gloucestershire (1998), but in the same year it vanished in the west of the country, on that chalky dumping ground in the far south-east a new colony was forming.

The new colony formed on those 111 acres of new English land because that Chunnel waste heap was designed to give something back to nature. In 1993, after creating artificial hillocks and lagoons, the bulldozers left Shakespeare Cliff Lower Construction Platform, and seeds of thirty-one species of native wild flowers were scattered across the bare chalk. The site remained closed to the public for four years, long enough for nature to poke around and decide whether or not to move in. During that time a local English teacher renamed the site Samphire Hoe after a line from Shakespeare's *King Lear*. 'There is a cliff whose high and bending head looks fearfully in the confined deep . . . Halfway down hangs one that gathers samphire.'[2]

Early Spider Orchid seed was not one of the species scattered, but the earth was unadulterated by chemicals, the climate perfect and competing plants absent so the area proved ideal for orchid seed blown from a nearby cliff-top colony. In 1998 sixty-one Early Spiders appeared on the Chunnel's spoil heaps. Within six years those sixty-one peppy little pioneers had become a super-colony of over 9,000.[3]

Since then, of their own volition, at least 170 more plant species have taken up residence at Samphire Hoe, among them Common Spotted and Pyramidal Orchids. As with all orchids, where the Early Spiders thrive, so does much of nature's tapestry. More than two hundred bird species stop off or make their home there, as well as thirty species of butterfly, 380 species of moth, thirteen dragonfly and damselfly species, and adders, common lizards and slow-worms. All this on virgin earth wrenched from beneath the sea.

To further enhance the area in a way sympathetic to the co-evolved tapestry of the land, small herds of livestock are occasionally brought in. There's no mowing of grass, no cut vegetation left to rot and change the soil's chemistry, no spraying of chemicals to control weeds. Here, as in nature, there are no weeds. Man working *with* nature prevents the kind of scrub invasion which plagues less artfully managed reserves, and that small new patch of England thrives.

By car, Samphire Hoe is reached by a dark, steep tunnel guarded by traffic lights. The first time you descend you get the distinct impression that you've accidentally entered a service tunnel to the Chunnel and the next time you see daylight – if ever – will be from France. Then daylight streams in from the end of the tunnel and you arrive at the Hoe. It's green and bright and, in mid-spring, blustery. There are information boards, a visitor centre, a chunky concrete sea wall, a car park, an odd, angular, blue, lighthouse-like tower, a concrete building housing an air-flow system for the Tunnel, and footpaths across the gentle contours of the engineered land. Three hundred feet-tall cliffs nuzzle the sky, but where vegetation has worn away to reveal bare chalk they look a little threadbare.

One day, straying off the designated path, I stepped carefully onto the short sward to tiptoe through hundreds of Early Spiders emerging from the grass. Most of the little plants scarcely reached the top of my hiking boots. They all seemed to face out to sea as if they were the arrayed regiments of a floral army, animated by the cold easterly, exuberant and stalwart, their little furry chocolate-and-lime heads shaking like defiant sentries. It was a marvellous, perplexing sight. What strange, unfathomable flowers they were. What a place it was where hope and action met. What a marvel that they could arrive in such numbers that they scorned their Schedule 8 status. Going extinct in county after county, there, properly managed, given a chance, they flourished *alongside* the concrete, machinery, people and cars.

There are no huge dumps of chalky earth near me, but there *are* areas where construction work has excavated foundations for houses, industrial units, car parks and supermarkets and formed piles of virgin earth. Some of those heaps get landscaped, turfed, planted with non-native species, mown and sprayed – a shame given what could be learned from Samphire Hoe. Others are abandoned, like empty burial mounds waiting to be blanketed in green. The Early Spider's success on the Chunnel's chalky waste gave me an idea for a new guerrilla strategy.

After a number of seasons gathering and propagating orchid seed, it became obvious that I had way more seed than I needed. A couple of Bee pods amounted to around 5,000 seeds; a couple of flower spikes' worth of Common Spotted probably over 200,000. If I planted all that on agar, I could end up with thousands of little orchids, which would be fantastic . . . if I had space to propagate them. I had quantities of orchid dust more appropriate to a small industrial enterprise than to a salad drawer, a shelf in a fridge and a small backyard.

If dried in the correct way, many species of orchid seed stay viable in the fridge for years, so that was where I put the seed I didn't have time or space to sow. Nevertheless, I kept collecting and storing the seed from my carefully nurtured orchid orchards. My seed surplus grew.

At some point, after the defeat of another batch of contaminated dishes, it occurred to me that instead of trying to maximise the germination rate of my orchid seeds by replicating natural levels of light and darkness, temperature changes, sowing periods, moisture and nutrients (with fungi and without), some of the job could, well, be done by nature.

Not every germinating seed needs petri dishes, diluted bleach and agar, of course. Lizards managed to appear on an Oxfordshire roadside, Tongue Orchids on the roof of a London City bank;

Bees mysteriously appeared in my dad's meadow and thousands of Early Spiders flourished at Samphire Hoe without so much as a blob of Malmgren's Medium. I decided not to leave my seed surplus sitting in the fridge. I might as well donate it back to nature.

That was when I started taking tubes of orchid dust out of the fridge and carrying them with me on the weekly shop and my daily commute, even on visits to the cinema, the pub, the car mechanic, the caterers for the wedding – not like very weird pets, but so that I could scatter seed in promising spots. If I spied a space, I'd check the coast was clear, scrape a shallow trench, pour a couple of hundred seeds into it, then cover them over with earth, pop the vial back in my pocket and walk away. It was much easier than planting tubers and rhizomes and could be done in broad daylight – for all anyone cared, I might have been assiduously stubbing out a cigarette.

Of course, different orchid species evolved to suit different habitats. That meant carrying different tubes of seed – Southern Marsh for damp areas, Early Purple for shadier ground, Bees for drier areas. Three different vials clinking around in my pocket wasn't ideal, so I settled on another idea: why not mix the seed from different species in one vial? That 'house' mixture became the one I carried – and carry to this day – a mix of different species to deploy where I can, allowing the seeds of the species most appropriate for that niche to develop and, with luck and time, flower, inspire and maybe even spread their seed back to the land.

Overall, seed-sowing is way more efficient in terms of time and resources, but with some species it seems to take longer for the plants to emerge – and I can't fuss over and protect them. As for its effectiveness . . . I'll always remember the moment a few years ago when Sylvia passed me her phone open on a Twitter feed alive with astonishment at the unexplained appearance of hundreds of Bee Orchids opposite a customer collection point at a shopping centre off the A30. I almost choked, astonished, flooded with elation and disbelief. I had scattered thousands of seeds there years earlier, but never expected they would thrive to that extent. Pride and joy pierced my heart. It was a bit like the time Nate said to me, 'That's OK, Daddy, you can borrow my dreams,' when I told him I hadn't dreamt the night before. On both occasions life let me know that we had helped something natural to occur, and it was wonderful.

19

BOG ANGELS

Summer can be cold on the moor. Autumn early. Winter lingers. So although it is mid-July, a cool breeze pummels Sylvia, Nate and me as we hike beside a meandering river along a narrow track beaten through heather by free-roaming sheep. We pass Heath Spotted Orchids, their speckled flowers browning around the edges. Lichen-splotched dry-stone walls. Skittish livestock. It takes a while to reach this little-known location off the tourist trail. At this point, near its origin, the river narrows to a pebble-bedded stream. Some of England's rarest orchids grow here. So remote is this bog that the breeze carries the rapid *popopop* of automatic rifles. The army's live-fire training grounds are just over the next brown hill. Although muted, the gunfire jars in this place. It is as if we have hiked into a war zone in one of England's national parks. Nate asks what the sound is.

'Soldiers practising.'

'Can we see?'

'We aren't allowed.'

'What are they practising?'

'Shooting things.'

'Why?'

'That's what soldiers do,' I say. 'Look,' I add, as a distraction, 'more sheep.'

Sheep, black-faced with yellowing horns like dirty croissants, dot the wide valley. They don't seem bothered by the rifle fire, just wary of the three apes intruding on their turf to find Bog Orchids.

First described as *Bifolium palustre* in John Parkinson's *Theatrum Botanicum* (1640), growing in 'divers places of Romney Marsh', Britain is home to around half the world population of Bog Orchids.[1] Since the late eighteenth century this species has lost its lowland habitat, while upland areas have seen numbers of livestock skyrocket, increasing the risk of overgrazing and trampling. The Bog Orchid is now extinct in most of England, including Romney Marshes. Where it can still be found, colonies that were hundreds strong now support only a dozen or so. In fact, it is possible that there are fewer of these plants in England than there are mountain gorillas in Rwanda.

It has been said that when the population of a species in a country falls below 500 individuals, that species is 'functionally extinct', for although some species – including orchids – can live on in reduced numbers for decades, it will only be for decades. Their populations are already presiding over an unavoidable total disappearance – unless humans intervene.[2] The population of Bog Orchids surviving in England today is probably around that fateful number (no one knows for sure), but despite their rarity and Britain's crucial position as their last great stronghold, no appeals to save them appear on TV. Their fate is not discussed in Parliament. No one of note seems to care. That is why we are here.

Recent genetic sequencing has placed Bog Orchids into their very own genus, *Hammarbya*, named after Carl Linnaeus's country house in Hammarby, Sweden, hitched to the species name *paludosa* (Latin for 'boggy'). Rather than going to the effort of renaming them, science might be better employed developing a Portable Bog Orchid Detector, because they are tricky plants to find. Lime-green, no taller than your thumb, they perch on mud or springy carpets of sphagnum moss at the margins of bogs where fresh water – 'flushes' – flows from springs. For unknown reasons, as they grow, the tiny green-yellow flowers twist on their stalks not just 180 degrees like most orchids, but 360, so that the flower's labellum and lower sepals are at the 'top' of the flower. This is where they would have been if no twisting occurred. In this position they form what appear to be two tiny angel wings.

Against the green-yellow of mosses, reeds and grasses, even when you know this is where you've seen them before, without a Portable Bog Orchid Detector, finding them is difficult. There is no guarantee they will be flowering; no guarantee that the cows haven't trampled them, a virus hasn't wiped them out or that this year's dry winter and unusually long, dry, cold spring hasn't shifted their flowering period by a couple of weeks, meaning that it might be too early to find them.

There are no Bog Orchids in our backyard. Their marginal habitat is too difficult for us to reproduce artificially. Today's mission is simply to see if we can find them. With luck they will be flowering. With even more luck, we can help them out with a little assisted pollination. In about eight weeks, we can return to collect and scatter their seed on suitable areas nearby. Another option with this particular species (it is one of very few British orchids which do this): if we are very careful, we may be able to tease off and replant one of the tiny *bulbils* (like miniature plantlets) which form on the edges of its leaves and can break off to become new plants.

Today we are the only people here. Dyed by peat, the gurgling stream is the colour of black tea. The gunfire has stopped. The rounded flanks of the hills form a wide amphitheatre carved by ice ages over hundreds of thousands of years. In the distance a crumbling tor scrapes the pale blue like a ruined tower. Beneath its gaze we crouch where water seeps through sphagnum moss near where beds of sundews lie. From a distance the brown hills look barren, but hiking through them, attuned to little things, life here is abundant. Crickets click. There is the *chuck chuck* of stonechats flitting from prickly gorse tips to granite blocks, to wind-stunted rowan. Nate peers at the sundews' sticky scarlet leaves splayed on the moss, captured insects slowly being digested. In the same spot, three years ago, I saw seven precious little plants.

We crouch and watch and search. Nate has seen photos of what to look for but is more interested in carnivorous plants. Here their population is healthy. Thick carpets of feathery yellow-green moss stretch between glinting pools where diminutive plumes of marsh grass shake wildly like the headdresses of angry little natives. We sink up to our knees in the marsh and squelch around to the next possible vantage point. The stream has eaten a deep fissure into the ground. Reeds obscure it. A wrong step and we'll fall in. Scrutinising every inch of ground before we step on it, careful not to crush anything out of the ordinary, we search. Wild ponies watch. We are their afternoon entertainment.

It's a small area and we progress around it slowly for three-quarters of an hour. By that point Nate is bored. Sylvia leads him to an accessible area of the stream so he can take off his boots and paddle. I spend a few minutes surveying the terrain, listening to Nate's gleeful commentary – 'Ooh! It's cold! Look, a dragonfly!' – and reconsidering our plan of attack.

My line of sight takes in the whole area. I don't want to admit it, but it may be time to accept that there are no orchids here. My heart numbs, not because of the time this search has taken or the wet feet, and not only because I wanted Nate to see them, but because I may be witnessing not just the loss of this small orchid, but an irreversible change to its habitat, this ecosystem, this land. I can't help but wonder what we are doing – really doing – here.

The last time circumstances made me reflect in a similar way was towards the end of the hot summer of 2020. To curb the spread of the Covid-19 virus, all non-essential shops and places of work were shut, people were only allowed to leave their house for one hour's exercise a day and the gates to playgrounds were padlocked. In parks, grass-cutting and grounds management ceased. Instead of being carefully mown and trimmed, public lawns, football pitches and roadside verges flourished. Wild flowers, finally unrestrained, stretched to their full height, blossoming, self-seeding. House sparrows, song thrushes, blackbirds, chaffinches, greenfinches, robins, red admirals, small tortoiseshells, gatekeepers and painted ladies emerged from who-knows-where in spectacular numbers, flitting, hopping, swooping and singing as if they could not believe the unimagined bounty of those rich new lakes of life.

Sylvia, Nate and I wandered carless roads to visit guerrilla-planted Bee Orchids blooming near a bus stop, while in the park the Common Spotteds spread pink, white, mauve and magenta columns of speckled flowers like a little forest of brash and beautiful beacons. We saw success. We saw hope.

We were not alone. Lockdown forced many people to abandon the home-commute-work-commute-home regime for a connection with outdoor space. The hour permitted for outdoor exercise became treasured, all the more so if it could be taken beneath the soft shade of trees. Some suggested that the pandemic had been caused in part by over-exploitation of the natural world: wildlife smuggling, the encroachment of people into areas where deadly viruses lurked – that and our habit of crowding together in urban spaces.

It was realised that years of sedentary indoor living combined with lung-weakening air pollution made humans more suscep-tible to Covid-19. Amid newsflashes of mass graves, exhausted medics, daily death and infection statistics came reports that world leaders were looking beyond the pandemic. They made announcements that seemed to signal a realisation that continued ransacking and poisoning of the natural world had to end – that nature's health was, after all, important to human health. It was time, said Britain's prime minister, Boris Johnson, to 'build back better'.[3] Before the first year of the pandemic was over he declared,

> We will use the UK's extraordinary powers of invention to repair the economic damage and build back better. Now is the time to plan for a green recovery . . . to make our nation cleaner, greener and more beautiful . . . we will harness Mother Nature's ability to absorb carbon by planting 30,000 hectares of trees a year by 2025, and restore the abundance of nature by rewilding 30,000 football pitches worth of countryside.[4]

The policy paper then goes on to reassure readers that this plan will protect 'future generations from climate change and the remorseless destruction of habitats'.[5]

Every restaurant, pub and café and most high-street shops in the country were shuttered. There were queues to enter supermarkets. Oil prices crashed. Petrol stations struggled to entice anyone to fill their tanks because only ambulances, joggers, walkers and cyclists used the roads. Britain's right-leaning Conservative government lurched left and paid the wages of those whose businesses and livelihoods had stalled. Perhaps we really had stepped through the looking glass. Perhaps the nation's leaders had woken up to the urgent need to coexist harmoniously with nature. Parks really might become rewilded. Roadside verges might be meadows. New houses might all have solar panels and grey water collection systems. Working from home, without queues of commuter traffic clogging the atmosphere with greenhouse gases and pollutants, might become the 'new normal'. Had habitual consumerist destruction grown up?

The first clue of what was really to come occurred within a week of Covid-19 restrictions easing. Nate and I were wading through one of those incidental park meadows in an area which was essentially grassed-over wasteland near some train tracks, without flower borders or shrubs, activity areas or goalposts. It was a place where dog walkers played catch with dogs. I had planted Common Spotted Orchids on a rise there years before and (blessings can always be mixed) was wondering whether all the long grass might smother them out of existence. While Nate and I identified butterflies, the grumbling of a diesel engine reached us.

Two council workers in high-vis orange vests were bumping down the wide footpath in a small flatbed truck. They stopped, stepped out and, like shock troops preparing for an assault, donned orange helmets, snapped down clear plastic visors, pulled on elbow-length gloves and hoisted petrol-powered strimmers onto their shoulders, then yanked them into life. Those hand-held steel-and-plastic monsters revved and crackled and ruthlessly mowed down the ranks of their foes. The air was smithereens and splinters and scraps of torn life amid blue farts of petrol fumes.

There are ways and times to cut meadows that are necessary for them to flourish, but that was a massacre. Helpless, I watched, rooted to the spot. Mesmerised by the shrill roar, Nate took my hand. Minutes of petrol clouds later, hundreds of square metres of meadow had been obliterated. The air was alive with confused insects. A few lucky birds picked them off. The flowers, the habitat, the magic were gone. So that was what the authorities meant by ceasing 'the remorseless destruction of habitats', making the nation 'greener and more beautiful' and restoring 'the abundance of nature'.

Nate tugged my hand.

'Is it over?'

'I think so.'

It made me consider what we were doing – *really* doing. Why should my family use our limited resources to shore up fragments of a doomed world when others get paid, using the tax *we* pay, to destroy it? Why was I up before dawn to guerrilla-plant orchids or search for them in development sites? Why was Sylvia scanning forums and planning applications so we could save threatened species, when people at the local council earned a living by rubber-stamping their demise? Why were we risking imprisonment and fines to protect and conserve a few species that have enchanted people for centuries? As long as there were councils with cutting regimes that wrecked thriving habitats, it seemed futile to even try to protect any living piece of natural heritage. So what were we doing?

Fast-forward to the present day and politicians' soundbites rarely evoke Mother Nature or 'building back better'. The precious daily hour spent outdoors has ceded its place to the more familiar and for many perhaps more satisfying rut of the daily commute, with a sprinkling of shopping, eating out and holidays abroad. Most of the lockdown joggers and cyclists have vanished; innumerable cars have returned.

The Intergovernmental Panel on Climate Change, the World Wide Fund for Nature, the United Nations, the International Monetary Fund, the World Health Organisation – every single institution and researcher with a remit for or affected by environmental issues – make dire predictions of a worsening climate crisis caused by human action and inaction. They talk of the approach of tipping points, of natural disasters causing trillions of pounds' worth of damage every year and, in the next thirty years, hundreds of millions fleeing their homes as heatwaves stoke vast wildfires and powerful hurricanes and floods kill tens of thousands of people. They warn of diseases spreading and mega-droughts destroying crops, dramatically increasing water scarcity for immense numbers of people and contributing to massive deprivation, which in turn will trigger armed conflict.[6] And this is without sparing a thought for the impact the same changes will have on the rest of the living world.

Yet that news struggles to vie for attention with stories of murder, miscarriages of justice, war and celebrity shenanigans. It seems that people are tired of reports about our suffering planet. These particular sources of bad news have grown stale; the media's need to feed a hunger for new, more alarming, stories has replaced them with others. That is how the human mind works.

Of course, not telling that kind of story or choosing not to listen to it does not make that future go away.

As the sun beats down on the moor and a dragonfly rattles past on glassy wings, it seems to me there is only one reason why this bog is no longer home to its little angels.

There are no crops for miles. There is no logical reason for herbicide, pesticide or fungicide spray-drift or leaching; no scrub encroachment; no noticeable overgrazing or trampling. There is no recent mining; no construction work for housing developments; no fire damage; no sign of diseased plants, theft or vandalism. On the surface of it, this bog appears much as it must have looked centuries ago, but I suspect this picture of a calm, nature-rich nook belies insidious annihilation.

Over the past few years, my own experience of nursing orchid refugees, young plants and orchid orchards in my backyard has seen predictable weather patterns deteriorate and extreme weather surge. At the Guerrilla Orchid HQ mild, dry winters are followed by late frosts, record rainfall, record heat, record storms. Almost daily I move the pots into shade or shelter, or water them to keep them and the fungi they need alive. No one is doing that for Bog Orchids on the moor.[7] The orchids, the fungi and the pollinators they depend on cannot switch on air conditioning when it gets too hot; they cannot get taken to a refuge centre or have supplies helicoptered in. They don't even have a representative in our parliament or legislation that adequately protects them from humans, let alone climate.

Since 2002, Britain has successively broken temperature records that might have been expected to stand for decades or centuries.[8] The cold winters and wet springs needed by the Ghost and Bird's-nest may already be passing into distant memory. For bees and flowers, crucial timings are being affected. A warming world means pollinators emerge days earlier per degree of warming (male bees appear nine days earlier, females fifteen), but flowers change their blooming times less predictably.[9] Under stable climate conditions over tens of thousands of years, male bees have emerged before females, so for a few days they devote their amorous attentions to unwittingly pollinating insectiform orchid flowers. Climate breakdown is sabotaging this complex game.

Pollinators aside, the marginal habitat required by Bogs is very sensitive to changes in rainfall and temperature: a fall of water level by a few centimetres will desiccate them; a rise of the same amount will cause them to rot. Every orchid species is just as sensitive to this type of change. Recent extended droughts have devastated orchid colonies.[10] In dry years plants may stay dormant, may not flower, may be weakened and become less resistant to disease; they may lose their mycorrhizal partners due to soil desiccation or their flowers may shrivel before they form seed. All of this affects subsequent generations.[11]

Extinction's corridors are found not only in city council offices and Whitehall; they are in our living rooms and kitchens. The government does not demand we drive certain vehicles, eat certain foods, buy particular goods, devour energy and resources at unsustainable rates, spray the earth with chemicals or pay money into pension funds which invest in destructive industries. Governments do not *make* us acolytes to a regime of consumption and waste. To a degree, it is our choice to do these things. This choice has contributed to the disappearance of the last orchids from this place.

Doing our bit by reusing plastic bags, recycling plastic, card and paper, choosing paper straws instead of plastic ones and buying an electric car is all very well, but it falls a long way short of what needs to be done, and these acts cannot mask the simple truth that orchids plainly reveal: the situation is dire and getting worse. A precipice is looming. The absence of the Bog angels is an indicator of a suffering land. As orchids have reminded me, we are nothing without our land.

The saga of this place, the moor, weaves a tapestry of this island's past. It has been shaped over periods beyond human comprehension by volcanoes, glaciers, rain, sun, wind, microbes, early plants, complex plants and the unbroken forests that root-split rock, died, rotted and formed earth and peat; shaped again by the mammals that grazed and the people who quarried, mined, slashed-and-burned trees, farmed, introduced animals, drained mires and bogs, dug peat and laid walls and roads. It is dotted with tumbledown granite buildings, the relics of human under-estimation, over-exploitation, ignorance, hubris. Those are the signs writ large on this place, and they offer important lessons.

When I think back to those first, fateful Bees I encountered in Dad's mini-meadow, I think of the healing of my spine. The spine and the Bees seem to be connected in a way – as in a myth I once studied, the myth of the Fisher King. The Fisher King is a cautionary tale. If the ruler of the land is sick, so is the land. Crops die, rivers run dry. The kingdom becomes a wasteland. If the land is healthy, so is the king. The land can heal the king and the king can heal the land, or they can cause each other's demise, for they are, in a sense, one and the same.

There is a lesson here for all of us who care about the future and the health of everything living in and on this planet. Within that lesson is the truth that though these extraordinary, enthralling plants spend months underground, orchids do not belong to a separate, different world. They are one of the wonders of *this* world, and we are a wonder too – the survival of both, and so much more, is interlinked in complex ways, the threads of a tapestry.

Wilks, Ash and others like me can guerrilla-save as many orchids as we can; we and the Sainsbury Orchid Conservation Project can propagate millions of plants, while scientists delve into the secret workings of every being; we can turn every park into a species-rich meadow, recycle plastic, drink coffee from bamboo cups – but without addressing the impact that our collective habits are having on all life forms (from soil-dwellers to plants, pollinators, ecosystems, food chains) and crucial planetary systems (water, carbon, nitrogen cycles and weather patterns), any effort to save the orchid-shaped thread of that infinitely complex web is like throwing another deckchair off the *Titanic*.

All the fascinating, complex modifications orchid flowers have undertaken over millions of years, all the adaptations that allowed them to outlive the previous mass extinction will no longer be enough to ensure their survival. Year by year, during the rest of my lifetime and throughout our son's, without urgent reform, losses like this will accelerate. That is the future forewarned by the absent Bog angels.

Faint bursts of machine-gun fire sound again. Comparing this national park to a war zone feels less like a metaphor. The war for the future is being fought all across the world without conventional weapons. The threat comes from invisible gases, synthetic chemicals, storm-force winds, rainfall, heat. It is a war we are waging against land left to us by ancestors who toiled here, fought here, died here, shaped this land and became it.

'Let's go,' I announce.

'Home?' asks Sylvia.

I shrug.

'Boots on,' Sylvia says to Nate.

He's having fun; he doesn't want to leave.

Happy to let him splash around a bit longer, I wait. A solitary skylark is singing somewhere, cascading frenetic jubilant notes. The wind passes. Sap runs through the world beneath my feet. The land exudes particles; I breathe them as the land feeds fungi, fungi feed plants, plants feed sheep, sheep feed people. Bacteria in the guts of worms help worms digest and form earth; sundews absorb insects through sticky leaves. Curled clumps of lichen on the granite boulders are each an ecosystem, cyanobacteria and fungi working together, harmonious.

This is a place of teeming life. This life overflows human bounds, quivers in the breeze, in the skylark's unjaded exuberance, the bracken dying back to feed the resting land. It is water running through the peat to the stream, the river; flowing over Nate's toes to the sea, to the sky, falling as rain; being part of his body or one day lapping against the ice sheets of Antarctica or an African beach. Millennia from now, some of those same molecules may fall here again, but will there be any Bog Orchids when they do?

Orchids were medicine, then goods, then scientific curiosities, puzzles to be solved, evolutionary clues, a challenge to the existence of an omnipotent Creator God. Then they were forgotten as people abandoned the land for factories, cities and machines. Britain's quietly vanishing orchids have become harbingers of what could be a global collapse. They remind us that we do not exist in a world separate from them or they from us. This is a warning from a future which, on this patch of moor, has already arrived.

I think about this as Sylvia dries Nate's feet on her top. I think about the fact that, for all we know, this is the only ball of rock in the incalculable vastness of the universe which supports any such rich, complex, curious, ancient, unknowable, beautiful tapestry of life. Thread by thread, our habits are unravelling it before we really know what it does.

And then I realise what it is we're doing here. Despite what sometimes feels like the misguided values of the modern world stacked against us, all we are doing, all we want, is what all parents want: to make this world the best place possible for Nate. That is what we have tried to do, one plant at a time: to act; to save a sliver of the fast-disappearing richness of this land; to leave it healthier, more bountiful, for all generations to come.

Holding hands, the three of us trudge back across the brown hills. The lark's song ends. Listen.

20

SILENT REVOLUTION

November. Predawn. A few lights burn in empty windows. A fox lopes past a faculty building. I walk quickly, cautiously, across the campus, carrying two supermarket bags containing lives that supermarkets do not sell: the four refugee Autumn Lady's-tresses I rescued weeks ago during a different dawn.

In this slightly undignified mode of transportation, each plant, nestled in a plastic pot, is a low rosette of leaves with a spiral flower spike still attached. The tiny pale petals have withered as ovaries swell with the precious dust of a new generation. I am moving them to what I hope will be a safer home on a south-facing bank of short turf beside a footpath at the edge of these university grounds. According to my pH meter the soil is neutral, a good place for these honey-scented orchids first described in English nearly five hundred years ago. Here they are: Gerard's 'Sweet Bollocks', a potent provocation of lust; a plant which fascinated Darwin with its complex pollination mechanism; the Near Threatened species that set me on a path of legally questionable action to save it from the same fate as Summer Lady's-tresses.

It has been a journey full of surprises, from there to here in the predawn dark. I have discovered a history of orchids stretching back over a thousand years which has shown ways in which they and humans have long been interlinked. People have prized them; studied, eaten and drunk them; collected them; assigned desires and morals to them; saved, abandoned and destroyed them. This is proof, were it needed, that we do not live in a world separate from nature. That is why wild orchids matter. They can teach us about humans, remind us about the past and warn us about what is coming.

The state of Britain's orchids has made me all too aware that, within my lifetime, our nation has lost much of its natural tapestry. In turn, this has made me realise that Nate was born into a world of accelerating extinction. We don't want devastation stalking his schooldays. We would rather he, like every child, grows up in a world of wide horizons, abundant, mysterious life, unpolluted water, undefiled air. That is why we do what we do, to protect a few threads of the world for him. We hope these threads might bind others tighter too. Then, one day, Sylvia and I can look him in the eye and say, 'We did what we could to save the wonders of this land for you.' Then we want to add, 'And it worked.'

Fortunately, the history of orchids also offers hope. Lady's Slipper, Monkey, Military, Irish Lady's-tresses and Ghost Orchids have all been declared extinct on these islands, only to reappear decades later. Projects such as Samphire Hoe and, dare I say it, our own successes show that nature can succeed. It just needs opportunity and a helping hand.

On my knees, on the bank, by the light of a torch, I dig four holes into the turf. Tenderly, I tip each plant into the earth and palm it in. Around them I rearrange some of the extracted soil, moss and grass, to obscure evidence that they have just been planted. I put the empty pots in the bag, wipe my earthy hands on the tousled pelt of cool, wet grass, switch off the torch and leave with that familiar sense of farewell.

I can do little more for those orchids now. But I can hope that next year they will flower again and maybe someone – a professor, student, cleaner, it matters little who – will be inspired by their spirals of white-green bells. Inspired enough, perhaps, to learn about the history and the fate of these flowers. If we can change their fate for the better, we can improve the future for our children and our land.

These are the last orchids I will be rescuing for a while. The micropropagation lab has been packed away. The crowded back-yard and pots have been emptied, most of the orchids hurriedly deployed. The Guerrilla Orchid HQ has been sold. Now Nate is bigger, we and the orchid refugees need more space.

Our new home, located deep in Devon's humpbacked hills, has no yard. Instead, there is a 50-metre-long strip of garden, rampant with nettles, bindweed and dandelions. I am already planning a meadow, a vast rare-orchid orchard that will produce the dust I can scatter across the country to boost their struggling populations.

At the end of the garden squats a dilapidated stone outhouse. Many years ago it was used for keeping chickens. We plan to convert it into a solar-powered orchid lab. Maybe I can save up for a laminar flow cabinet.

The future of guerrilla orchids will differ in other ways too. Conscious of the environmental footprint associated with our reintroductions, we want to reduce our use of plastic, focus on spreading seed in key locations reached by foot, bike, public transport.

The species I aim to propagate will only be those most at risk and which have a good chance of being successfully re-introduced in local areas. By focusing on two or three such species we hope to streamline the propagation process, allowing me to experiment with the best possible germination techniques, times and growing media for these particular orchids. Ultimately I aim to reintroduce thousands of these plants. I want them to become common again.

Given that it can take years for orchids to flower, I know the next phase of the mission will take time. I accept that for what it is: nature's timescales, seasons and cycles, not humankind's sometimes headlong rush. Orchids have taught me to savour time. They have also taught me it does not require governments, policies or laws to make a difference. All it needs is the will to act, to do what is right – whatever the potential cost – to bring little reminders that the living world is a place of wonder back to areas of concrete and brick.

Sometimes, of course, crowds bustle past orchids we have planted without a glance, anxious to reach somewhere else or distracted by phone screens – perhaps even reading someone else's post about the very plants they are passing. They hurry by as if the secrets to life are waiting at their destinations or can be found on their phones – eternally elsewhere, in a video, vlog, blog, meme, reel or post. I too once thought important secrets could be found far away. I travelled and travelled, restless, to find them, but it was Britain's orchids and the land where I was raised that showed me the most important answers lay right here. I had simply been too distracted to see them.

Half a plump moon watches as I retreat from the campus. That billions-of-years-old satellite has witnessed the rise of continents and their separation, the creation of an atmosphere and its poisoning, the emergence of life and its extinctions. It will watch over these plants, all the others I have guerrilla-planted and all those we have yet to send out into the world. It will watch over them and Nate long after we have fed back into the great tapestry from which we all emerge for our brief turn beneath the sun, and it will soon, I hope, see changes taking place, nature given space, our battered lands helped to heal.

Swinging the bag of empty pots, I walk down the carless road. Birds in nearby trees start to sing. Sunrise is coming.

Orchis mascula

Early Purple Orchid, from W. Curtis, *Flora Londinensis*, vol. 2 (1778).

APPENDIX: BRITAIN AND IRELAND'S WILD ORCHIDS

Over the years I have gained insight into many of Britain's wild orchids. However, my first-hand experience of saving them is limited to those present in my part of the country. I have not saved orchids in Wales, Ireland, Scotland or northern England. As such, the orchids mentioned in this book are a small percentage of the variety still to be found (if you are lucky) in parts of Great Britain and Ireland. The species near you may differ considerably from those near me, but wherever you are, and whatever species you find, nationally and internationally, orchid species are in decline.

Below are Great Britain and Ireland's generally recognised native orchids.[1] A few species, such as Tongue Orchids, are considered recently naturalised; I have not included them because the extent to which they are now 'native' is not yet clear. I have listed the species in the order of their usual flowering periods (which can shift, earlier or later, depending on climate and location), together with their conservation status as of 2019, approximate historical loss and information on whether current legislation in England and Wales offers them special protection.

Many of Britain's orchids, including some uncommon ones, have a conservation status of Least Concern. Researchers emphasise that 'Least Concern' does not mean that species is of no conservation concern, only that it is not currently threatened with imminent extinction.[2] On the other hand, species falling under the categories Critically Endangered, Endangered, Vulnerable and Near Threatened are considered at risk of imminent extinction. Policy (where there is any) is drawn to the most threatened species; meanwhile, those considered of Least Concern slowly become rarer.

The following information was collated from two recommended books: Anne and Simon Harrap's *Orchids of Britain and Ireland: A Field and Site Guide* (2nd edn, 2009) and Sean Cole and Mike Waller's *Britain's Orchids: A Field Guide to the Orchids of Great Britain and Ireland* (2020).[3] The names and distribution of orchids in these recent books make an interesting comparison with those in earlier classics like Colonel Godfery's *Monograph and Iconograph of Native British Orchidaceae* (1933), *The Wild Orchids of Britain* by the novelist (and orchidophile) Jocelyn Brooke (1950) and *Wild Orchids of Britain* by V. S. Summerhayes (1951).

Early Spider Orchid (*Ophrys sphegodes*). March–May. These little insect mimics can grow in colonies of hundreds or thousands, often on chalky south-coast cliff tops. Usually no more than 20 cm tall, they have several brown, fuzzy flowers. Slightly angled to the sky, these blooms cling to a lime-green stem, looking less like spiders and more like little paper models of strange lime-and-chocolate lollipops quivering in the breeze. In the 1950s they were recorded along much of England's south coast; that continuous band has since become fragmented and inland colonies have not fared well. However, in contrast, some robust coastal colonies have seen plant numbers increase and new colonies (Samphire Hoe is one – see Chapter 18) have appeared. Red List Species Status (2019): Least Concern. Lost from 73 per cent of its former range. Extinct in Wiltshire (1989) and Gloucestershire (1998). Schedule 8 protection.

Early Purple Orchid (*Orchis mascula*). March–June. Colonies of these flowers emerge like regal purple hyacinths early in the year, growing among bluebells, primroses and cowslips on old hedge-banks and unsprayed verges, in churchyards, ancient woodlands and meadows. Where I live, their appearance marks the start of orchid season. If it's not too hot and dry, their flowers last several weeks. Their flower spikes can be 50–60 cm tall and covered with dozens of flowers ranging from salmon-pink to rich purple. Shakespeare's *Hamlet* (act 4, scene 7) describes 'long purples' in the garlands Ophelia wears when she commits suicide by drowning. Debate shifts back and forth about whether these 'long purples' really are Early Purples or not. If they are, they are Shakespeare's only explicit reference to orchids; if they aren't, it would seem strange that Shakespeare, whose work is replete with references to the natural world, does not include at least one reference to the well-known aphrodisiacs of his day. Early Purples can be found in diverse habitats and soils and are widespread across the United Kingdom, but colonies are now often small and localised – a pattern that is increasingly common with orchids and increasingly concerning. While the species numbers may not be low enough to warrant a more threat-ened conservation status, such colonies are likely to lack genetic diversity, a key component for the long-term survival of most popula-tions. Red List Species Status (2019): Least Concern. Recently lost from 28 per cent of its historic range. No special protection.

Green-winged Orchid (*Anacamptis morio*). Mid-March–late May. Often considered one of the most beautiful of Britain's orchids, they can be found blooming in the vicinity of Early Purples and Early Spiders. Each flower usually has green-veined 'wings' or a bonnet (formed by the flower's upper petals), cupped around the flower's column. The lip is often speckled white and pink, but there can be a significant range of colours within a single colony – some plants have white flowers, others ranging through soft pinks to rich purples. It looks spectacular in large numbers on the short turf of a properly managed meadow, petals fluttering on a gusty May day. As its favourite habitat of unimproved pastureland has become increasingly rare, so has this once very common orchid. Sadly emblematic of this decline is the fact that the Green-winged grew in abundance around Darwin's house and was one of the inspirations for his book on orchids, but, since then, it has completely vanished from the area. Red List Species Status (2019): Near Threatened. Lost from 50 per cent of its previous range in around fifty years. No special protection.

Dense-flowered Orchid (*Neotinea maculata*). Early May–mid-May. A small orchid which self-pollinates. Its small, pale yellow/green flowers are, as the name suggests, densely packed and it can look a bit like a diminutive torch of pale fire. Found in patches of western Ireland and Northern Ireland; now extinct in Britain (once found on the Isle of Man). Near Threatened status in Ireland, where it has lost 46 per cent of its previous range since its discovery in 1864.

Fly Orchid (*Ophrys insectifera*). Mid-May–late May. An exquisite example of insect mimicry, despite its name, the Fly is designed to lure male digger wasps. Spaced along a tall, thin, straight stem, the 1–2 cm-long flowers have a deep reddish-brown labellum with a lighter coloured band (*speculum*) which resembles (to wasps) sunlight reflecting off the folded wings of another wasp. Their three tapering sepals are lime-green and their two upper petals are reduced to the size and appearance of antennae. Against a leafy backdrop they can be tricky to spot, but when your eye is in, many of these flowers can suddenly materialise into view like strings of splendid gems. As recently as the 1950s, the Fly was found throughout most of southern England, along the Welsh Marches, through most of the Midlands, the north-east, and parts of Scotland; it has since vanished from many of those areas. Red List Species Status (2019): Near Threatened. Lost from 58 per cent of its previous range. Extinct in Devon, the Isle of Wight, Middlesex, Essex, Norfolk, Cambridge, Huntingdonshire, Worcestershire, Herefordshire, Shropshire, Staffordshire, Leicestershire, Lincolnshire and County Durham. No special protection.

Loose-flowered/Jersey Orchid (*Anacamptis laxiflora*). Mid-May–late May. Looking a bit like the Early Purple Orchid (although its leaves are narrow, plain green and held erect, very different from the blotched, strappy leaves of Early Purples, which tend to hug the ground), it looks like a string of narrow pink/purple and white blooms loosely arrayed around an upright stem which is 15–60 cm tall. Only found in the Channel Islands and where naturalised by Kew at Wakehurst. Lost from at least 25 per cent of its former range on Jersey. No special protection.

Lady Orchid (*Orchis purpurea*). Mid-May–early June. This showstopper of an orchid is most often found in Kent (in the past, it had a wider range across the south-east). A stately orchid with people-shaped flowers:

its sepals and petals form a dark-red bonnet, while the lobes of the broad labellum make 'arms' and the lady's reddish spotted and speckled 'skirt'. Sometimes a metre tall, each plant can have fifty flowers gathered like a cloud of tiny skirted women whisked into the air on a sudden breeze. There can be quite a lot of colour variation within a single colony, with some of the skirts and bonnets yellowish and green, others deep purple. It can be found on well-drained alkaline soil in open pastures and dappled shady glades. Red List Species Status (2019): Vulnerable. Lost from 57 per cent of its previous range. No special protection.

Man Orchid (*Orchis anthropophora*). Mid-May–early June. First described in English by John Ray in 1690, as the name suggests, this orchid possesses slim, yellow-green humanoid flowers. Each outlined in pencil-thin ruby red, the little figures dangle from an oversized 'cap/ helmet' formed by the tightly curved upper petals and sepals. It likes calcareous earth, mostly in south-east England, where these green spires, shaggy with little people-shaped blooms, can be found in the increasingly scarce areas untainted by chemicals. Red List Species Status (2019): Endangered. Although lost from nearly 60 per cent of its previous range and now extinct in Somerset, the Isle of Wight, Hertfordshire, Buckinghamshire, Cambridgeshire, Leicestershire and Derbyshire, new colonies have appeared in recent years and, in Britain, it could benefit from climate change. No special protection.

Monkey Orchid (*Orchis simia*). Mid-May–early June. Previously thought extinct on several occasions. Early records of this now very rare orchid often confuse it with the Military and Lady, so it is diffi-cult to assess exactly which species is being described and how far its range has diminished. Now its jumbled clouds of slender-limbed monkey-shaped blooms are publicly accessible in only two reserves in the UK, one in Oxfordshire, one in Kent. Red List Species Status (2019): Vulnerable. Lost from at least 80 per cent of its previous range. Schedule 8 protection.

Military Orchid (*Orchis militaris*). Mid-May–early June. Believed to be extinct for decades, the rare and majestic Military presents pink-white flowers with darker purple spots that look like wounded soldiers;

as Gerard wrote, 'little flowers resembling a little man, having a helmet upon his head, his hands and legs cut off'. Growing to about 45 cm tall, a colony of Military on display in a woodland clearing is a spectacular sight. Red List Species Status (2019): Vulnerable. Lost from 84 per cent of its previous range. Now only found in three sites. Schedule 8 protection.

Sword-leaved Helleborine (*Cephalanthera longifolia*). Mid-May–early June. While other orchids are quirky or flashy, tiny or weirdly humanoid, the graceful *Cephalanthera* helleborines possess class and elegance. Some, like the Sword-leaved (named after its glossy green, long, thin pointed leaves), prefer woodland margins – glades, clearings, the edges of footpaths and bridleways – rather than sunny situations. The Sword-leaved has pure white flowers with a rich golden patch on their labella. Even when fully open, the blooms look half-closed, as if they are just too languorous to offer themselves fully to the world. Red List Species Status (2019): Vulnerable. Has 'suffered a dramatic decline over the last 200 years'.[4] Lost from 74 per cent of its previous range. Now extinct in Devon, Dorset, Wiltshire, East Sussex, Kent, Oxfordshire, Norfolk, Herefordshire, Shropshire, Derbyshire, Nottinghamshire, Lancashire, Yorkshire, County Durham, Northumberland and seven counties in Scotland and Wales. No special protection.

Bird's-nest Orchid (*Neottia nidus-avis*). Mid-May–mid-June. Known to early herbalists as a 'degenerat' kind of orchid, the Bird's-nest is named after its clump of twiggy, bird's nest-like rhizomes. One of Britain's three achlorophyllous natives, it lives almost entirely underground in the leaf litter of damp and gloomy woodland (usually beech or hazel), gaining nutrients via a tripartite relationship with soil fungus and nearby trees. The only visible parts of the orchid are the old-bone-coloured flowers, which are pollinated by a range of insects. Darwin's experiments suggested that if an insect touches 'pressure pads' on the flower's rostellum, the flower releases a drop of gluey liquid. The pollinia stick to this substance and get carried to flowers on the same or a different plant to effect pollination. If the flower isn't pollinated within a few days, the pollinia can fall apart and crumble onto the stigma, allowing self-pollination.[5] The Bird's-nest is now known also to reproduce vegetatively via its underground rhizomes.[6] Distribution maps from the 1950s show its presence across almost all of Britain, but maps from 2020 show its existence has become much patchier, probably as a result of habitat loss and climate change (it needs the right damp conditions to flower). Red List Species Status (2019): Near Threatened. Lost from at least 54 per cent of its historical range. No special protection.

Burnt/Burnt-tip Orchid (*Neotinea ustulata*). Mid-May–mid-June (another *forma* seems to flower later: early–mid-July). The flower spike of a Burnt Orchid is an exquisite dense little cloud of petals, deep burgundy, crimson and ivory white. I liken them to black-cherry-sauce-dipped frilly gourmet desserts. Each sweetly scented flower is the shape of a tiny man, his 'head' buried in a red cap. Burnt Orchids usually flower for several years at a time, then either die or go dormant underground, where the fungal partner probably plays an important role in nurturing the tuber, but little is known about this relationship. They tend to like south-facing slopes on calcareous soil. Where these conditions exist I have found them (like Frog Orchids) growing on the ramparts of ancient hill forts. Books on native orchids from the 1950s show its distribution stretching across vast swathes of England, from South Devon to Cumbria. Since then it has vanished from almost everywhere. Red List Species Status (2019): Endangered. Lost from 79 per cent of its previous range. Now extinct in Devon, North Somerset, West Kent, Surrey, North Essex, Middlesex, Buckinghamshire, West Suffolk, West Norfolk, Cambridgeshire, Huntingdonshire, Northamptonshire, West Gloucestershire, Herefordshire, Staffordshire, Shropshire, Leicestershire, Rutland, Nottinghamshire, West Lancashire, South-east Yorkshire, South-west Yorkshire, South Northumberland, Westmoreland, Cumberland. No special protection.

White Helleborine (*Cephalanthera damasonium*). Mid-May–mid-June. In some ways physically similar to the Sword-leaved, the White can be found rising from carpets of fallen beech leaves in gloomy wood-land in southern England, its relationship with soil fungus enabling it to grow where few other plants get enough light to thrive. Its creamy white flowers are usually even more tightly shut than the

Sword-leaved, but through the narrow opening a rich golden centre can be seen, earning it the name 'the poached egg plant'. As Darwin discovered, the flower doesn't need to open because it is usually self-pollinating. Red List Species Status (2019): Vulnerable. Lost from 41.5 per cent of its previous range. No special protection.

Coralroot (*Corallorhiza trifida*). Mid-May–late June. The Coralroot gets its name from its rhizomes, which resemble cream-coloured, knobbly coral growths. Found in circumpolar regions of the world, in the UK it only grows in eastern Scotland and a few areas of Northumberland and Cumbria. The county flower of Fife, this small orchid lives entirely underground, where, like the Bird's-nest and Ghost, it gets the nutrients it requires via a tripartite parasitic relationship with fungi and nearby trees. About 10 cm tall, the slender flower spike and its little forest of self-pollinating yellow-green and white flowers are the only parts which emerge above ground. Red List Species Status (2019): Vulnerable. Lost from 46 per cent of its previous range. No special protection.

Common Spotted Orchid (*Dactylorhiza fuchsii*). Mid-May–mid-July. The Common Spotted belongs to the same genus as the Marsh Orchids and has many physical similarities. It has paler flowers (more pink and white than red or purple) and its labellum displays dark loops or lines, rather than only specks. It can be locally common where the earth is sufficiently moist. A range of insects has been found carrying Common Spotted Orchid pollinia, but some studies show that long-horn beetles are the main pollinators. This is a widespread species which I tend to encounter in its largest numbers in woodland clearings, but in the same habitats I have often found colonies succumbing to the invasion of bracken and bramble. Red List Species Status (2019): Least Concern. Widespread but recently lost from 14 per cent of its previous range. No special protection.

Common Twayblade (*Neottia ovata*). Mid-May–mid-July. One of the first orchids ever recorded in English (in 1548 in William Turner's *The Names of Herbes*, where it is referred to as the 'Martagon'). Since then, as often happens with orchids, its name has shifted, to 'Twayblade' (also Tway-blade, Twaiblade and *Herbe Bifoile*). Parkinson's *Theatrum Botanicum* (1640) referred to it as the 'Ordinary Wood Bifoile', 'Twayblade' and '*Bifolium sylvestre vulgare*'; Curtis's *Flora Londinensis* applied the scientific name '*Ophrys ovata*'; Trimen and Dyer's *Flora of Middlesex* (1869) called it '*Listera ovata*'. In the early 2000s, molecular phylogenetics (research into genetic heritage) discovered that all orchids in the genus *Listera* descend from an ancestor common to the *Neottia* orchids. Because the *Neottia* genus name is older, in taxonomic terms, that meant all orchids in the *Listera* genus were relocated to *Neottia*. That is why, today, the Common Twayblade is also *Neottia ovata* – *Neottia* from Ancient Greek for 'nest' (its tangled, twig-like rhizomatous roots said to resemble an untidy bird's nest); *ovata* from Latin for 'egg-shaped', referring to the shape of its leaves, of which it usually has only two. These twin leaves are the basis for its common name: 'tway' for 'two'; 'blade' from *blatt*, German for 'leaf'. Red List Species Status (2019): Least Concern. Lost from 27.5 per cent of its historical range; 'a relatively large proportion of the losses being recent'.[7] No special protection.

Irish Marsh Orchid (*Dactylorhiza occidentalis*). Late May–mid-June. Today, science recognises five species of Marsh Orchid in Britain – the Southern Marsh, Early Marsh, Northern Marsh, Irish Marsh, Pugsley's (or Narrow-leaved) Marsh, plus three species in the same genus without the name 'marsh' (Common Spotted, Frog and Heath Spotted). There are also a lot of subspecies and varieties. Debate continues over whether there might still be previously unidentified species (or are they simply varieties?).[8] As its name suggests, the Irish Marsh is only found in Ireland. Its flower spike, more compact than other Marsh species, carries ten to twenty rich purple flowers. Red List Species Status (2019): Least Concern. First considered a separate species in 1930, since then lost from 18 per cent of its range.

Lady's Slipper (*Cypripedium calceolus*). Late May–mid-June. A flagship orchid species once thought extinct, it tends to be rather shorter in real life than it appears in pictures, but is no less spectacular. Each of its enchanting flowers has a distended labellum which forms a faintly obscene yellow pouch (the 'slipper') used to trap solitary bees for pollination. A victim of its beauty, this species was collected to near complete extinction in Britain during the nineteenth century. Aside from genetic issues (discussed in Chapter 15), attempts to reintroduce it have been hampered by the fact that, like people, slugs and snails appear to be irresistibly drawn to its flowers. Red List Species Status (2019): Critically Endangered. Harrap and Harrap place the historical range loss at 95 per cent, but with only one acknowledged 'wild' plant persisting in Yorkshire this percentage is probably rather low.[9] Schedule 8 protection.

Late Spider Orchid (*Ophrys fuciflora*). Late May–mid-June. Not discovered until 1828 in England, the Late Spider is among the rarest of Britain's orchids. Dubious records from the nineteenth century suggest it was once found in Surrey, Dorset and Suffolk, but now it is only found in very localised areas in Kent (where it was once more widespread). A close relative of the Bee, it too is an insect mimic, but its two to six flowers have an extra set of 'horns' between the sepals, and its labellum is broader and differently marked from the Bee. Unfortunately, its usual bee pollinator does not exist in Britain. However, as very occasional hybrids between the Late Spider and the Early Spider and Bee have been found, it suggests that, at least sporadically, an insect takes pollinia from one and delivers them to another. Red List Species Status (2019): Vulnerable. Very localised in south-east England. Lost from 33.5 per cent of its previous (very limited) range. Has been a target of recent thefts from the wild. Schedule 8 protection.

Lesser Twayblade (*Neottia cordata*). Late May–mid-June. Being small and slender with tiny, fine, reddish-green flowers, the Lesser Twayblade looks very different from the Common Twayblade, but its pollination mechanism is the same, as are its twinned leaves. It is rare in England (more common in Scotland and around Snowdonia); its favoured habitat is damp, cool heath and moor, where it can be found on north-facing slopes on damp sphagnum moss, often under heather and blueberry plants. Red List Species Status (2019): Least Concern. Lost from 44.5 per cent of its previous range. Now extinct in Shropshire, Derbyshire, Cheshire, Flintshire, South-east and South-west Yorkshire and the Isle of Man. No special protection.

Pugsley's Marsh Orchid (*Dactylorhiza traunsteinerioides*). Late May–end of June. A more delicate-looking Marsh Orchid, its flowers are usually purple-pink, its labellum marked with the darker speckles, streaks and spots characteristic of Marsh Orchids. It is more specific in its habitat requirements than some other Marsh species: it likes very wet fens and flushes, usually alongside black bog-rush. Red List Species Status (2019): Least Concern. Recently lost from 26 per cent of its previous range. Now extinct in Northumberland. No special protection.

Greater Butterfly Orchid (*Platanthera chlorantha*). Late May–early July. With blooms looking not much like butterflies, the Butterfly Orchids are striking, ethereal plants, especially by moon- and starlight when they also emit a lily-like fragrance to attract large, night-flying moth pollinators. The Greater Butterfly is reasonably widespread and can be found in a range of conditions (woodland edges, pastures, hedge-banks, railway sidings), but colonies are increasingly small and localised. Red List Species Status (2019): Near Threatened. Lost from 46 per cent of its historical range. No special protection.

Early Marsh Orchid (*Dactylorhiza incarnata*). Late May–early July. Widespread but localised, a single colony of Early Marsh can display a range of colour variations (from creamy white, through pinks, crimsons and purples) and size (from squat plants of a few centimetres in height to showy specimens over 50 cm tall). It has narrower flowers than the Northern and Southern Marsh and can be distinguished by the stuck-up 'ears' formed by its lateral sepals. Despite its name, I have found it in not especially wet unimproved meadows and in dune slacks. Red List Species Status (2019): Least Concern. This species has several subspecies, some of which are very rare. Lost from 43.5 per cent of its previous range. No special protection.

Heath Spotted Orchid (*Dactylorhiza maculata*). Late May–mid-July. Another widespread member of the Marsh Orchid genus, this is closely related to the Common Spotted; the flower has a broader labellum and looks like a wide, frilly petticoat. Unlike many orchids, it prefers slightly more acidic soil conditions, as found in moorland, mires and peat. Red List Species Status (2019): Least Concern. Lost from 21.5 per cent of its previous range. No special protection.

Bee Orchid (*Ophrys apifera*). Early June–early July. Often featured as the exotic-looking insect-mimic poster-flower for Britain's orchids, its bee-like nuggets set in their three-pointed pale amethyst sepals are always an uplifting sight. Despite its mystique, it's quite widespread and can be found in a variety of habitats across England and parts of Wales. It behaves like a 'weed' as an early coloniser of disturbed ground, but it will disappear if the surrounding vegetation gets too long. Unfortunately, one good flowering season is not always followed by another, as Bees can get exhausted from flowering then setting seed, so they sometimes stay dormant the following season. Several variations on the standard Bee exist, each with different shapes, markings and colours; var. *chlorantha*, for example, is entirely pale green and white. Red List Species Status (2019): Least Concern. Lost from 16.5 per cent of its previous range. No special protection.

Northern Marsh Orchid (*Dactylorhiza purpurella*). Early June–mid-July. Generally, where the Southern Marsh Orchid's traditional territory ends, the Northern Marsh takes over. After the Southern Marsh was recognised as a distinct species in 1914, the Northern Marsh was designated a species in 1920. It tends to have more magenta-crimson colouring than the Southern Marsh's purple-pink flowers, and its labellum has a distinctive diamond shape. Red List Species Status (2019): Least Concern. Lost from 21 per cent of its previous range, many losses recent. No special protection.

Hebridean Marsh Orchid (*Dactylorhiza ebudensis*). June. The most localised of Britain's Marsh Orchids, only found in the Outer Hebrides, where its squat, rich purple flowers appear in good numbers in the machair. It is not always given species status by orchidologists even though, along with the Common Spotted, it is the ancestral parent species of three other Marsh Orchids (Northern, Southern and Pugsley's) and, with the Heath Spotted, the ancestral parent of Irish Marsh. No Red Book Status. Loss of previous range unknown as it was not considered a separate species until 1976. Currently considered part of the Pugsley's Marsh Orchid complex.

Southern Marsh Orchid (*Dactylorhiza praetermissa*). June. The Southern Marsh has a stout stem and a spike of closely packed flowers, which can be pale pink to deep purple. Given the right moist conditions, it can create pink-to-purple carpets across riverside meadows and wetter areas of dune slacks. Until recently not found north of a line from the Ribble to the Humber (beyond that was the territory of the Northern Marsh) but, with climate change, this species is rapidly spreading further north. Red List Species Status (2019): Least Concern. Classified as a separate species in 1914, since then lost from 20 per cent of its previous range. No special protection.

Common/Chalk Fragrant Orchid (*Gymnadenia conopsea*). June–July. Three Fragrant Orchids exist in Britain (they have been variously combined and separated as species over the past couple of centuries). The Common Fragrant grows in dry, species-rich pasture and hay meadows. Its flowers are small and pink, and each has a long nectar spur protruding downwards from the back of the flower. Their dozens of little blooms are held in tall, slender spires, from which they emit a cloying sweet aroma. White variations also exist. Red List Species Status (2019): Least Concern. Lost from 39.5 per cent of its historical range. No special protection.

Pyramidal Orchid (*Anacamptis pyramidalis*). June–mid-July. The Pyramidal can be found offering its bright pompoms in meadows, dune slacks and downs generally below a line drawn from Devon to south Yorkshire. When opening, it has a distinctive pyramidal shape, but when all the flowers are open the small, colourful, densely packed flowers form more of a column, sphere or cylindrical shape. They can range in colour from white, through delicate coral pinks, carmine, indigo and lilac, and can also vary in size and height. Red List Species Status (2019): Least Concern. Lost from 20 per cent of its previous range. No special protection.

Lizard Orchid (*Himantoglossum hircinum*). Mid-June–late June. Probably the weirdest-looking wild flower in Britain, its flower spikes can be a metre tall, festooned with the tassels of lizard-tail-like labella. Lizard populations ebb and flow through history, with a constant population existing for over two hundred years near Dartford, Kent, until 1850, when, probably due to over-collection, that population died out. Until 1900, plants appeared sporadically in Kent, Suffolk and Hampshire, but during the first half of the twentieth century the Lizard started to make a comeback, growing in numbers and increasing its range. The reason for this could be warmer, wetter winters, fewer frosts, drier and cooler summers and autumns – conditions more closely matching the climate of western France, where the Lizard is reasonably common. With Lizards increasingly appearing in haphazard locations, this appears to be a climate-change winner.[10] Its most consistent population has long been associated with Sandwich Bay, Kent, where it can appear in thousands on golf courses and front lawns. Red List Species Status (2019): Least Concern. Lost from 82.5 per cent of its previous range. Schedule 8 protection.

Musk Orchid (*Herminium monorchis*). Mid-June–early July. This is a tiny orchid. Its flower spikes, seldom more more than 10 cm tall, are smothered in small, bell-shaped, frilled, pale-yellow blooms. Despite its name, it smells of honey. The name 'monorchis' means 'one-testicle' because these plants have tiny, single, round tubers no bigger than a small ball-bearing. Their tubers grow close to the surface and, being small, do not have much storage capacity. As a result the Musk can be significantly affected by drought: unfortunate given that dry, hot springs and summers are becoming more frequent. Increasingly rare, I have only found them on steep chalky banks on ancient common land. Red List Species Status (2019): Vulnerable. Lost from about 69 per cent of its previous range. Now extinct in Somerset, Glamorganshire, Oxfordshire, Rutland, Cambridgeshire, Suffolk, West Norfolk. No special protection.

Fen Orchid (*Liparis loeselii*). Mid-June–mid-July. First noted by John Ray in 1660 in Cambridgeshire, the spindly green-yellow, skywards-facing flowers are often difficult to spot. An orchid oddity in Britain, the Fen is essentially a tropical orchid (it has a pseudobulb, not an underground tuber or rhizome) which somehow manages to survive in our temperate climate, although only in fens and dune slacks. Very rare and declining until recently, the Fen needs a specific marshy habitat and ground disturbance to survive. Red List Species Status (2019): Endangered. Lost from 63 per cent of its previous range but concerted conservation efforts by the charity Plantlife have resulted in a successful reintroduction programme in Suffolk and Cambridgeshire. With correct management, numbers in South Wales and the Norfolk Broads have also reversed a steady decline.[11] Schedule 8 protection.

Frog Orchid (*Dactylorhiza viridis*). Mid-June–mid-July. Described in the seventeenth century as growing in many places in southern Britain, the Frog is another Marsh Orchid, but would never be confused for others in the genus because its flowers look very different. They are more humanoid (but without arms) and are apple-green, sometimes with deep russet tones. The Frog doesn't really look much like a frog until you imagine a frog in mid-leap, hind legs trailing behind, front legs either side of its head as it propels itself forwards into the air. Unlike other Marsh Orchids, this one doesn't just look as if it produces nectar, it actually does – and a variety of insects come to visit it. When they do, pollinia get stuck to them and the pollinia take twenty minutes to complete their forward arc, ready to pollinate a different plant (in comparison, Early Purple pollinia arc forward in a few seconds). Although early accounts talk about its noticeable presence in southern counties, and as recently as the 1950s it is described as common and widespread, it is now only relatively common in Scotland and increasingly rare elsewhere, mostly as a result of modern commercial farming techniques. Red List Species Status (2019): Vulnerable. Lost from 60.5 per cent of its previous range. Now extinct in Essex, Norfolk, Middlesex, Huntingdonshire, Warwickshire, Lancashire, Monmouthshire, Montgomeryshire, the Isle of Man, Dumfriesshire, Wigtownshire, Berwickshire, Dunbartonshire and Kent. Found in one site each in Suffolk, Cambridgeshire, Lincolnshire, Herefordshire. No special protection.

Heath Fragrant Orchid (*Gymnadenia borealis*). Mid-June–mid-July. Looking similar to the other Fragrant Orchids, it is not so dependent on alkaline earth and likes more acidic heath and grassland, often in western and southern parts of Britain and Ireland, though commonest in Scotland. Possibly Britain and Ireland's only endemic orchid, exactly what its range is – or has been – is difficult to say as it has only been considered a separate species since 1988. Its scent is a beautiful mix of honey and cloves. Red List Species Status (2019): Least Concern. Harrap and Harrap state that there 'has undoubtedly been a considerable decline [in numbers]'.[12] No special protection.

Lesser Butterfly Orchid (*Platanthera bifolia*). Mid-June–mid-July. The daintier of the Butterfly Orchids, the only sure way to tell it apart from its sibling is to take a good look at the shape of the flower's column and the angle of its pollinia (the Lesser Butterfly has a slim column with the pollinia parallel; the Greater has a fatter column, the pollinia coming together in an upside down V). In 1951 Victor Summerhayes gives its distribution as across Great Britain and Ireland; the startling decline experienced by this species is shown when this is compared with the distribution given by Cole and Waller in 2020. In sixty-nine years the elegant Lesser Butterfly has vanished from hundreds of square miles across the Midlands, northern England, East Anglia, central Scotland, southern and western Ireland. Red List Species Status (2019): Vulnerable. Lost from 64 per cent of its previous range. No special protection.

Small White Orchid (*Pseudorchis albida*). Mid-June–mid-July. First recorded on Mount Snowdon by John Ray in 1670, as its name suggests this is a small orchid with white flowers. The flowers are small too – smaller than its ovaries – and form a drooping cluster on an upright, lime-green stem which is usually 10–20 cm tall. Like Creeping Lady's-tresses and Lesser Twayblades, this orchid prefers a cooler climate, which may be one factor why, in recent years, it has disappeared from southern England. Red List Species Status (2019): Vulnerable. Lost from over 65 per cent of its previous range. Now extinct in Kent, Sussex, Gloucestershire, Herefordshire, Staffordshire, Shropshire, Derbyshire, Cheshire, Lancashire, South-west and North-east Yorkshire, Monmouthshire, Glamorganshire, Carmarthenshire, Cardiganshire, Montgomeryshire, Flintshire and much of Ireland. No special protection.

Dark-red Helleborine (*Epipactis atrorubens*). Late June–late July. First described by Ray in 1677, this species grows in open rocky places – limestone outcrops, old quarries, cliff ledges and rocky slopes – in only a few areas of northern England, Scotland, Wales and western Ireland. *Epipactis* helleborines have straight, more robust flower spikes than the graceful *Cephalanthera* helleborines. Their flower's labellum forms a noticeable deep cup (the *hypochile*), which can contain sugary liquid attractive to pollinators. The petals and sepals are generally more like blunt-tipped, five-pointed stars drooping from the stem. The upright flower spikes of *Epipactis* can routinely be over 50 cm tall and (compared to *Cephalanthera*) usually quite smothered in blooms, so they can resemble shaggy spires. The Dark-red's flowers have a bright yellow anther which contrasts with the rich claret of the rest of the flower. Red List Species Status (2019): Least Concern. Lost from 30 per cent of its historical range. Extinct in Breconshire and Denbighshire; may have once been found in Gloucestershire and Herefordshire. No special protection.

Red Helleborine (*Cephalanthera rubra*). Late June–early July. More pink in colour than red, this is one of the UK's rarest flowers. First recorded in 1797, it appears to have once been reasonably common in the Cotswolds, with colonies also found in Somerset and West Gloucestershire. Like other *Cephalanthera* helleborines, it has a slender, graceful stalk, which in this case rises to about 40 cm in height. When its lilac-pink flowers open wide, the lateral sepals are outstretched like welcoming wings and the upper sepals and petals form a loose hood over the paler centre of the flower. Although the Red Helleborine does not produce nectar, research suggests that, to its pollinators (solitary bees), it looks blue, so tricks them into resembling the nectar-producing *Campanula*. Research also suggests that it engages in a tripartite relationship with trees and fungi. Frustratingly for conservationists, it is a very exacting species in terms of the correct amounts of light and shade required for it to flower. Today the Red Helleborine exists in Britain as only a handful of plants in fewer than five guarded locations. Red List Species Status (2019): Critically Endangered. Lost from at least 70 per cent of its previous range. Schedule 8 protection.

Lindisfarne Helleborine (*Epipactis sancta*). Early July–mid-July. Only found on the drier parts of dunes on the island of Lindisfarne in Northumberland, this pretty, tall, green-and-white-flowered orchid was recognised as a separate species in 2002, but then reconsidered as a subspecies (at best) of the Dune Helleborine. At present its species status is looking increasingly tenuous. Red List Species Status (2019): Endangered. No special protection.

Creeping Lady's-tresses (*Goodyera repens*). July. Looking similar to other Lady's-tresses (even though it belongs to a different genus), this dainty orchid has small, puckered, white, sweet-smelling flowers blossoming off a green stem. The flowers in this instance are characteristically hairy. It was first noted in the 1770s in Scotland – Scottish pine forests remain its stronghold. An enigmatic exception are some colonies, first discovered in 1885, in Norfolk. These were probably introduced when Scots pines were transported to the area to create plantations. Ecology attitudes to these Norfolk colonies have shifted over the years from studious conservation to regarding them as an 'introduction', leading to the removal of the SSSI status of the site where many grow. Red List Species Status (2019): Least Concern. Lost from 44 per cent of its previous range. Now extinct in Dumfriesshire, Peeblesshire, Orkney, Yorkshire. No special protection.

Marsh Fragrant Orchid (*Gymnadenia densiflora*). July. Its historical loss of range is unclear as this was only recently considered a separate species (it had been regarded as a separate species in the early nineteenth century, then merged with other Fragrant Orchids, before several studies separated it again between the 1980s and 2000s).[13] The Marsh Fragrant has a tall spire smothered in small, pale lilac flowers which I have seen perched on slightly raised areas of wet dune slacks. It also favours disused quarries and pristine fens. Its scent is one of Sylvia's favourites: a divine blend of honey and cloves. Red List Species Status (2019): Least Concern. No special protection.

Marsh Helleborine (*Epipactis palustris*). July. First described in English in 1633, its ivory, damask and yellow flowers hang languidly from their stems. The fragrant Marsh Helleborine has a large, slightly frilly lip, which Darwin noted was hinged, so he suggested that when an insect landed, the lip moved downwards, then, as the insect backed out, the hinged lip rose again, forcing the insect upwards so the pollinia could stick to it;[14] other researchers suggest the hinged movement destabilises the insect so they raise their heads to keep balance, and this is sufficient for the pollinia to reach the insect. Each flower spike is about 20 to 40 cm tall. In the places where they occur, they can be present in large numbers, carpeting the damp area (often in dune slacks) and filling the air with a scent of vanilla. Red List Species Status (2019): Least Concern. Lost from 60 per cent of its previous range. Now extinct in the Channel Islands, Bedfordshire, Huntingdonshire, Northamptonshire, Worcestershire, Radnorshire, Dumfries and Galloway, Roxburghshire, Berwickshire, the Lothians and Fife. No special protection.

Dune Helleborine (*Epipactis dunensis*). Mid-July–late July. Knee-high, often a dull green at first glance, closer inspection shows these to be surprisingly colourful, with bright-yellow anthers, pouting lips washed with violet and deep-mauve hypochiles (the 'cup' part of the *Epipactis* flower). Like other helleborines, this species has caused a century of headaches for taxonomists: it was first recognised as a subspecies in 1921, then received species status in 1926, was demoted in the late 1970s to a variety, then DNA work in 2002 re-elevated it to a species. Because of this confusion, the exact number of Dune helleborines in the world is uncertain. Red List Species Status (2019): Insufficient Data. For the reasons given above, loss from its previous range is impossible to accurately calculate. No special protection.

Narrow-lipped Helleborine (*Epipactis leptochila*). July–early August. Found in dim areas of ancient deciduous woodland growing on calcareous soil, its subtle green-and-white blooms seem to cascade from a tall, straight stem. Largely self-pollinated. Given the shady conditions it needs, it is probable this orchid forms a close relationship with soil mycorrhizae and may have a tripartite parasitic relationship with nearby trees. For a long time it was confused with Broad-leaved and Green Helleborines and different helleborine subspecies, so its true range was unclear; however 'it is now clear that Narrow-lipped

Helleborine is far less widespread than previously thought'.[15] Red List Species Status (2019): Insufficient Data. Lost from 50 per cent of its previous range. Extinct in Devon, West Sussex, Monmouthshire, Kent and Hertfordshire. No special protection.

Bog Orchid (*Hammarbya paludosa*). July–August. Usually between 4 and 8 cm tall, this is Britain's smallest orchid. Its tiny, dingy-green flowers are difficult to spot and it has very specific habitat requirements (perched on mud or moss in bogs which are continually wet and with a constant flow of water). This makes it vulnerable to the hot, dry summers and irregular rainfall caused by climate change. Unusually, its flowers twist 360 degrees, so they are the 'right' way up. Today, it is more frequent in western Scotland than elsewhere. Red List Species Status (2019): Least Concern. Lost from 61 per cent of its previous range. Britain holds about 50 per cent of the world's population. Now extinct in Kent, Sussex, Surrey, Hertfordshire, Bedfordshire, Cambridgeshire, Suffolk, Lincolnshire, Staffordshire, Cheshire, Lancashire, County Durham. No special protection.

Green-flowered Helleborine (*Epipactis phyllanthes*). Mid-July–early August. A lover of damp woodland and willowy riversides, as the name suggests this orchid is all green. Its flowers are often quite small, green with a white or yellow lip, and they hang down like pendants, essentially facing the ground. A self-pollinated orchid, it can appear in many slightly different forms (with permutations of leaf shape and size, and how open the flowers are), leading to the naming of many different varieties. Red List Species Status (2019): Least Concern. Lost from 36 per cent of its previous range. Now extinct in the Isle of Wight, Somerset, Warwickshire, Derbyshire. No special protection.

Broad-leaved Helleborine (*Epipactis helleborine*). Mid-July–mid-August. A widespread orchid, often found at woodland edges, sometimes by car parks in woods, on railway sidings, along the edges of shady lanes, it seems to prefer the company of certain tree species (hazel and beech). It can be quite variable in appearance, from small and delicate to tall and robust, with a flower spike up to a metre tall, and its droopy, slightly dishevelled-looking flowers can range from light green to purple-red. When in bud, the stem arcs over like a whip and gradually straightens, the flowers (for impatient orchid lovers) seemingly taking for ever to open. Red List Species Status (2019): Least Concern. Lost from 31 per cent of its historical range. No special protection.

Irish Lady's-tresses (*Spiranthes romanzoffiana*). Mid-July–mid-August. Like other Lady's-tresses, this orchid has small, white, sweet-smelling flowers, but in this species they are ranked in three spirally columns and the plant itself has grass-like leaf blades (other Lady's-tresses have ground-hugging rosettes of compact leaves). Up to 20 cm tall, its little spirals rise out of damp grassland. As the name suggests, this is more of an Irish species, but it can also be found in the west of mainland Scotland and the Hebrides. Until 1993 a colony existed on the edge of Dartmoor. Its disappearance was thought to indicate this orchid's extinction in England and Wales, then in 2019 a colony was discovered in Wales. Red List Species Status (2019): Least Concern. Lost from 19 per cent of its previous range. Britain and Ireland hold the only European populations. Extinct in England since 1993. No special protection.

Ghost Orchid (*Epipogium aphyllum*). Mid-July–mid-September. Once upon a time referred to as the Spurred Coral Root, what is now known as 'the Ghost' can be found in damp woodland in parts of continental Europe (but rarely in large numbers). England seems to be the limit of its westward range, and the flower spikes here tend to be considerably smaller than on continental specimens. It has become the holy grail of orchid lovers, with dedicated search parties venturing out to find it. For many years (so far) their missions have not met with success; in recent times searching has therefore been scaled back, as it was realised that many feet tramping through fragile habitat where this orchid could be growing under the surface was probably detrimental to the plant. Achlorophyllous, the banana-smelling Ghost exists under the damp leaf litter of ancient woodland, where it gains nutrients from soil mycorrhizae and nearby trees. In order to flower it needs damp springs and cool conditions, so that the leaf litter never dries out – conditions which are increasingly unpredictable. Decades have passed between sightings. Red List Species Status (2019): Critically Endangered. Last seen in 2009. Extinct in Britain? Schedule 8 protection.

Violet Helleborine (*Epipactis purpurata*). Late July–mid-August. First written about in 1807, this is among the last orchids to flower every year. It likes shady woodland where its clusters of pale green-ivory flowers hanging from a greyish-purple stem, open quite wide, clearly displaying their five-pointed-star shape (the flowers don't droop as much as on some other helleborines). It is a long-lived plant and, according to Harrap and Harrap, it may not flower until it is thirty years old, with larger plants living for centuries.[16] Red List Species Status (2019): Least Concern. Lost from 38.5 per cent of its historical range. Now extinct in Devon, Cambridgeshire, Norfolk, South-west Yorkshire. No special protection.

Autumn Lady's-tresses (*Spiranthes spiralis*). Mid-August–October. The smallest of the Lady's-tresses at about 10 to 15cm tall, this is one of the first orchids described in English (in 1548 in William Turner's *The Names of Herbes*) and it is the last orchid to flower in the year. Its leaf rosettes appear in the autumn, continue throughout the winter and (often, but not always) die away in May or June. The flower spike emerges from the centre in autumn. By the time the single spiral of ten to twenty little pale flowers develops, the old leaves have withered and a new rosette appeared next to, but separate from, the flower spike. I have seen the graceful little spires blooming as late as November, their flowers unfazed by a covering of frost. In the 1950s it was common throughout most of England and Wales; now it is only likely to be encountered in southern England and Ireland in areas of short dryish turf where, if the conditions are right, it can form colonies of hundreds, filling the air with honey scent. Red List Species Status (2019): Near Threatened. Lost from 55 per cent of its historical range. Almost gone from northern and eastern England and the Midlands. Currently extinct in Warwickshire, Staffordshire, Nottinghamshire, Derbyshire, Cheshire, Middlesex. No special protection.

Lizard Orchid, from J. Smith and J. Sowerby, *English Botany*, vol. 1 (1790).

ACKNOWLEDGEMENTS

This book exists because of the support and encouragement of many to whom I owe heartfelt thanks:

I am incredibly grateful to Patrick Walsh, who encouraged me to shape the story of my orchid exploits, and to the rest of the team (in particular John Ash and Rebecca Sandell) at PEW Literary.

I have had amazing assistance from my publishers, John Murray. My special gratitude goes to Georgina Laycock, who passionately backed this project from the early days, and to Caroline Westmore (meticulous attention to detail), Sara Marafini (genius of book design), Susan Spratt (production), Alice Graham (marketing), Emma Mitchell (publicity), Katharine Morris (all-round assistance), Tim Waller (copy-editor extraordinaire), Howard Davies (scrupulous proofreading) and Kirsty Howarth (legal advice).

I am grateful to Paul Redshaw (insider information on the Lady's Slipper situation), Sean Cole, and to many others who prefer to remain nameless.

Thank you to my parents, John and Di, whose passion for plants and travel rubbed off on me; to Nela (protector of lives great and small); and most of all to my wife, Sylwia, whose idea it was that I write this book. Without your support, assistance and encouragement, your patience with my scribblings, kitchen-lab tinkerings and dawn departures, there would be no orchid outlaw. I love you more than words can say. Finally, but not in the least lastly, heaps of love and thanks to Nathaniel. Your boundless curiosity about the world is the perfect reminder that we should all be doing everything we can to support, conserve and save its wonder.

NOTES

Chapter 1: Quietly Vanishing

1. Natural England, 'Protected Plants, Fungi and Lichens', n.p.
2. Bardgett, pp. 6–16; Oost and Bakker, pp. 301–3.
3. Bardgett, p. 17, states: 'as a general rule, the average rate of soil formation, over the entire surface of the Earth, is around 10 centimetres for every thousand years, or just 0.1 millimetre a year.' Some soils form faster; some soils are millions of years old.
4. Cole and Waller, p. 232, refer to the intentional destruction of a site where the very rare Military Orchid grew.

Chapter 2: Orchid Magic

1. Watson, p. 521, says there were seven; Boyle, p. 60, states there were only three plants in Europe, all belonging to the baron.
2. Boyle, pp. 61–3.
3. Watson, p. 521.
4. Endersby, pp. 114–16.
5. Reinikka, pp. 138–9.
6. Darwin, *Orchids*, p. 279. Here Darwin refers to the work of renowned English orchidologist and taxonomist, John Lindley.
7. Schiff, p. 49.
8. 'Orchidaceae [are] the most threatened group of flowering plants worldwide' (Wraith and Pickering, p. 8).

9. Harrap and Harrap, p. 416; Cole and Waller, pp. 274–6.

10. Hinsley et al., p. 437, state that annual exports of potted orchids from the Netherlands alone were valued at around 500 million euros (in 2015) and that between 2006 and 2015 nearly 660 million live orchid plants were exported legally worldwide from the main orchid-exporting nations, Thailand, Taiwan, the Netherlands, Japan. This was an increase of 200 million orchid plants compared to 1996–2005.

11. Nong, pp. 25, 35, 38, 63, 96. Orchids continue to play a significant role in traditional Chinese medicine (Pei and Yang, pp. 1–3). See also Koopowitz, pp. 66–77, and Teoh, pp. 55–68. Teoh (pp. 10–11, 305–62, 363–6) notes that the continued use of orchids in medicine, food, cosmetics and traditional charms remains a contributing factor in their over-collection in many parts of the world.

12. Schiff, p. 2: 'The orchids grow in the woods and they let out their fragrance even if there is no one around to appreciate it. Likewise, men of noble character will not let poverty deter their will to be guided by high principles and morals' (trans. A. Poon).

13. Endersby, pp. 25–6, 57–9; Schiff, pp. 23–7; Teoh, pp. 109–16, 124–5.

14. Multiple texts consider Victorian Orchid Mania; this section draws upon some of them: Endersby; Reinikka; Griffiths; Hansen; Willis and Fry; Schiff; Teoh; Anghelescu et al.

15. Reinikka, pp. 219–22.

16. Endersby, pp. 108–9. Chapters 4 and 6 of Endersby's book offer an excellent analysis of Victorian Orchid Mania.

17. Schiff, p. 11.

18. For example, Albert Millican's memoir of orchid hunting (p. 151) relates how he felled 4,000 trees during one trip to South America to collect 10,000 plants of a single orchid species.

Chapter 3: Lessons of a Bee

1. Among the first of these new illustrated botanical reference books, the *New Kreüterbuch* by Leonhart Fuchs came out of Germany in 1543. Others soon followed, many of them based on previous works. One of these was by the Flemish physician Rembert Dodoens,

whose *Cruydeboeck* (1554 and 1563) supplemented (and in other ways simply plagiarised) Fuchs's book. Dodoens's book quickly became the most translated text after the Bible. Its English edition – the *New Herball* – formed the basis for Gerard's *Herball*. For fuller accounts of the botanical renaissance see Teoh, pp. 17–39 and Endersby, pp. 25–42.

2. Gerard (1597), p. 1383.

3. Ibid., pp. 1391–2.

4. Ibid., pp. 156–76.

5. Ibid., p. 161.

6. Ibid., pp. 156, 159, 166, 168.

7. Ibid., p. 158.

8. Endersby, pp. 26–7.

9. For more discussion and examples of the Doctrine of Signatures see Pearn, pp. 99–106; Endersby, pp. 18–23, 34–43; Schiff, p. 3; Richardson, pp. 9–11.

10. Theophrastus was probably influenced by the 'rhizotomoi', experts in the uses of roots, none of whose writings, if they produced any, now exist. Endersby, pp. 14, 18.

11. As quoted in Endersby, p. 38.

12. A century later, Nicholas Culpeper, in the book originally published as *The English Physitian*, observed that orchid roots 'are hot and moist in operation, under the dominion of dame Venus, and provoke lust exceedingly', and as a result 'are to be used with some discretion' (Culpeper, p. 240).

13. For more on salep/saloop see Teoh, pp. 13–17, 25; Endersby, p. 14; Koopowitz, pp. 67–8; Schiff, pp. 3–5.

14. Johnson, *Descriptio*, p. 130.

15. Parkinson, pp. 1345 and 1356.

16. Ray, *Historia*, pp. 1212–26.

17. As with other early writers, Blackstone's naming seldom matches today's, so whether these are fourteen distinct species by modern reckoning is uncertain.

18. Curtis, vol. 1, plate 66, accompanying text.

19. Ibid.

20. Ibid.

21. Ibid.; Gerard (1633), p. 219.

22. Typical of this is Henry Wood's late Victorian *A Season Among the Wild Flowers*, in which he explains the origins of the term 'Orchid' thus: 'The name, which is Greek, alludes to the tubers of the roots, especially those of the genus *Orchis*, whose plants have either *ovoid*, solid egg-shaped root-knobs, or *palmate* ones, i.e. shaped like a hand' (Wood, p. 112).

23. John Fowles wrote eloquently about these ideas in *The Tree* (1979).

24. Fowles, *The Tree*, p. 71.

25. Endersby, pp. 105–28.

26. Among the best known are John Lindley's *Sertum Orchidaceum: A Wreath of the Most Beautiful Orchidaceous Flowers* (1838) and *Folio Orchidacea: An Enumeration of the Known Species of Orchids* (1852–5); the Belgian Jean Jules Linden's *Lindenia: Iconography of Orchids* (1891–97); and the German Heinrich Gustav Reichenbach's *Xenia Orchidacea* (1858–1900). Society's focus on exotic species continued into the twentieth century. The title of Frederick Boyle's *The Woodlands Orchids, Described and Illustrated, with Stories of Orchid-Collecting* (1901) to me evokes images of native orchids in English woods, but his book is entirely about seeking tropical orchids overseas. In the United States, like Europe home to hundreds of native species, James O'Brien's *Orchids* (1911), L. Sherman Adams's 1943 title of the same name and dozens more are entirely about cultivating tropical species. Only in Australia did the situation differ. Many books were published about native Australian orchids, but this was probably because Australia had been sufficiently recently colonised that, for Europeans, its flora retained an aura of the primitive exotic.

27. Smith and Sowerby, vol. 6, plate 383, accompanying text.

28. Ibid., vol. 1, plate 34, accompanying text.

29. Darwin, *Orchids*, p. 25.

30. Trimen and Dyer, pp. 268–73.

31. Curtis, vol. 6, plate 67, accompanying text.

32. I return to the story of the Military Orchid in Chapter 16.

Chapter 4: Lost Lady's-tresses

1. Harrap and Harrap, p. 78.
2. Ibid., pp. 77, 83.
3. Much further along my journey of orchid discovery I wondered whether this might be the same area Charles Darwin refers to as 'near Torquay', where he examined dozens of Bee Orchids 'some time after flowering season', and Autumn Lady's-tresses too (Darwin, *Orchids*, pp. 54 and 113).
4. Gerard (1597), pp. 326–7.
5. Rasmussen, p. 199. Rasmussen (p. 268) cites others who suggest it could take 7–15 years after germination for Common Twayblades to flower for the first time.
6. Jacquemyn et al., 'Mycorrhizal Diversity', p. 3271; Rasmussen, p. 199.
7. Harrap and Harrap, p. 66.
8. Cocker's *Our Place* provides an excellent analysis of the (in)effectiveness of some of these institutions.
9. The Narrow-lipped Helleborine disappeared before 1930; White Helleborine, Lizard and Fly between 1930 and 1949.
10. Vidal (n.p.); Reid et al., pp. 8–9, 12–16, 25–35. Compare also Rackham, p. 97, who noted: 'As much ancient woodland was destroyed in 28 years [1950s–1970s] as in the previous 400 years.'
11. IUCN, 'Peatland Programme'; Morelle (n.p.) and Barkham (n.p.) both refer to the loss of English meadows since the early twentieth century as 97 per cent, a figure which has been widely repeated since.
12. Joint Nature Conservancy Council, 'UK BAP Priority List Species', n.p. The 2007 version is the most recent list (the system was devolved to different administrative regions soon after). As there is no longer a 'British' list, using this data to track the state of species across the UK is more complicated than it could be.
13. Ibid.
14. Riley (n.p.), for example, suggests over four hundred species have gone extinct in Britain since 1814.
15. DEFRA (2020), *UK Biodiversity Indicators 2020*, p. 32.
16. Joint Nature Conservancy Council, 'UK BAP Priority List Species', n.p.

17. UK Biodiversity Action Plan.

18. Environmental Performance Index.

19. Davis, 'UK has "Led the World"'.

20. Rackham, pp. 25–30; Zayed and Loft, pp. 6–10, confirm that the second half of the 1900s saw significant overall increases in most agricultural areas (orchards are a notable exception).

21. Ironically, excessive ploughing, fertiliser and pesticide use degrades soil health and makes crops *more* susceptible to disease (Monbiot, *Regenesis*, p. 20).

22. Duffy et al., pp. 969, 975; Kauth et al., p. 388.

23. Monbiot, *Regenesis*, pp. 64–5.

24. Manning, pp. 257–67; Monbiot, *Regenesis*, pp. 69–76.

25. In many ways, like the large extinct mammals before them, humans took over the shaping and management of the land: see Jepson and Blythe, pp. 11–41; Macdonald, pp. 13–19, 30–1, 87–101; Monbiot, *Feral*, pp. 90–120; Vera, pp. 51–60.

26. Rackham, p. 108; Vera, pp. 111–18, 123–88; Harrap and Harrap, pp. 13–14.

27. More information on the challenges faced by orchids available in Harrap and Harrap, pp. 12–18; Cole and Waller, pp. 278–80.

28. Reid et al., pp. 15–16, 26–35; Raum, pp. 2715–18.

29. Rackham, p. 48. Sumption and Flowerdew (pp. 156–8) note that there were short-lived gains in appearance of some orchids when rabbit numbers drastically declined, but that within three to four years grass species and shrubs dominated. They also note that reduced rabbit numbers (due to myxomatosis) had ripple effects across ecosystems, benefiting some species but not others, and being directly responsible for the extinction of Britain's large blue butterfly (p. 160).

30. Harrap and Harrap, pp. 16–17.

31. In IUCN, p. 159, it is admitted that there is 'considerable uncertainty' around population numbers of many species and that the classifications inevitably include a lot of estimation. One orchid expert (Koopowitz, p. 159), referring to the IUCN list of extinct plant species, notes that, in relation to tropical orchids, the IUCN categorizations bear 'little resemblance to reality and the true extent around extinctions'.

Chapter 5: Extinction's Corridors

1. The Young's Helleborine was removed from Schedule 8 in 2011, shortly after I encountered the Act; to date the Lapland Marsh Orchid remains.
2. MHCLG, National Planning Policy Framework, pp. 5, 52.
3. Rackham, p. 194.
4. Ibid., p. 52.
5. Macdonald, p. 78.
6. Rackham, p. 29.
7. Woolf, p. 99.
8. Rackham once noted, 'Tree planting is not synonymous with conservation; it is an admission that conservation has failed' (p. 29). He also observed, after describing the landscape as being like a historic library, that, 'Every year a thousand volumes are taken at random by people who cannot read them, and sold for the value of their parchment. A thousand more are restored by amateur bookbinders who discard the ancient bindings, trim off the margins, and throw away the leaves that they consider damaged or indecent. The gaps in the shelves are filled either with bad paperback novels or with handsomely-printed pamphlets containing meaningless jumbles of letters' (p. 30). Rackham was writing in the mid-1980s. Much of what he wrote then is just as relevant today.
9. United Nations, p. 142.
10. Harrap and Harrap, p. 19.

Chapter 6: Guerrilla Rewilding

1. Curtis, vol. 2, plate 62, accompanying text.
2. Turner, p. 70; Harrap and Harrap, p. 315.
3. Sometimes a single development can destroy thousands of plants. See Horton.
4. WWF, p. 4.
5. See Jepson and Blythe, pp. 45–85, for a practical overview of rewilding and accounts of several rewilding projects.

6. See for example Isabella Tree's account, *Wilding*, and Derek Gow's *Bringing Back the Beaver*.

7. See Monbiot, *Feral*, pp. 90–120; Jepson and Blythe, pp. 70–5, 137–42. The first European bison to be present in England for centuries were released in July 2022 into woodland near Canterbury, Kent, as part of a rewilding project.

8. A few headline examples: 'Council Mows Down Rare Orchids on Roundabout Despite 11-year-old Girl's Pleas to Let Them Grow to Support Bees' (Stringer); 'Rare Orchids Mown Down but Council Learns Lesson' (Scardarella); 'Bulldozers Destroy Thousands of Wild Orchids in Wolds' (Catton); 'Council Accidentally Bulldozes Roadside Nature Reserve' (BirdGuides).

9. Wild Flower Society, p. 4. Since I began guerrilla orchid rewilding the Wild Flower Society has replaced its code of practice with a different version of the BSBI guidance which simply states that 'introducing or planting out native plants can be unwise'.

10. Norris et al., p. 5.

11. Stroh et al., pp. 13–15.

12. The government's advisory body on conservation, the JNCC, concedes that plant ranges change and 'in many cases the (formerly clear) distinctions between "presumed-native" and "presumed-alien" ranges are becoming increasingly blurred' (Taylor et al., p. 5).

13. For notes on the chestnut and sycamore see Woolf, pp. 72 and 191 (these are not Britain's only introduced tree species); see Richardson for notes on betony (p. 34) and comfrey (p. 79) (these are not Britain's only introduced flowers); for discussion of scientific classification of archaeophytes and neophytes (naturalised and introduced) plants, see Stroh et al., p. 3.

14. Rackham, pp. 47–51; Monbiot, *Feral*, p. 70.

15. Why terrestrial orchids respond well to calcium is unclear, particularly as orchid seed grown in a laboratory seems to dislike its presence.

16. Gerard (1597), p. 158.

17. The renowned herbalist Nicholas Culpeper described it thus: 'These roots alter every year by course, when the one riseth and waxeth full, the other waxeth lank, and perisheth' (Culpeper, p. 240).

18. There is considerable variation in the sequencing of how different orchid species develop roots, shoots and tubers/rhizomes. See Rasmussen, pp. 200–8.

Chapter 8: Curious Contrivances

1. Parkinson notes that: '*Tragus* was of a strange conceit about the increase of these kindes of orchides . . . he thought that seeing they were not procreated by their own seede . . . they were increased by forraign seed, namely of Blacke birdes and Thrushes, that in their copulations let fall some of their sperme upon the ground' (p. 1346).

2. Quote from Yam and Arditti, p. 3. For more on the early 'science' of orchid seed, see ibid., pp. 2–3; Arditti, 'Discoverers and Firsts', pp. 815–17; Endersby, pp. 41–2. For a commentary on Darwin's orchid interest and publication of his orchid book, see Yam, Arditti and Cameron, pp. 2129–35.

3. Darwin to A. G. More, 2 June 1861, in Darwin, *Correspondence* 9, p. 159. In a letter to J. D. Hooker composed between 28 July and 10 August 1861, Darwin confided, 'It is mere virtue which makes me *not* wish to examine more orchids; for I like it far better than writing about varieties of Cocks & Hens & Ducks', ibid., p. 223

4. Darwin to W. D. Fox, 8 July 1861, in ibid., p. 196.

5. 'My interest in it [the topic of orchid fertilisation] was greatly enhanced by having procured and read in November 1841, through the advice of Robert Brown, a copy of C. K. Sprengel's wonderful book, *Das entdeckte Geheimnis der Natur*' (Darwin, *Autobiography*, p. 127). Darwin refers to the influence of both these works throughout *Orchids*.

6. Darwin to J. D. Hooker, 17 July 1861, in Darwin, *Correspondence* 9, p. 205.

7. Darwin, *Correspondence* 8, pp. 235–7.

8. Darwin to J. D. Hooker, 27 July 1861, in Darwin, *Correspondence* 9, p. 220.

9. Darwin to A. G. More, 7 July 1861, in ibid., p. 194; Darwin to A. G. More, 17 July 1861, in Darwin, *Correspondence* 9, p. 206.

10. George Gordon to John Balfour, 1 July 1861, in Darwin, *Correspondence* 13, pp. 445–6.

11. Darwin to William Darwin, 9 May 1861, in Darwin, *Correspondence* 9, p. 122.

12. Darwin, *Orchids*, pp. 67, 61.

13. Ibid., p. 1.

14. Ibid., pp. 1–2.

15. Edens-Meier and Bernhardt, pp. 5–6.

16. In a letter to A. G. More, 8 March 1861 Darwin admits, 'I have got many orchids in my garden including the rare Goodyera [repens],' *Correspondence* 9, pp. 49–50. *Goodyera repens* (Creeping Lady's-tresses) is not naturally found in southern England so it is highly likely that Darwin had a hand in introducing it, and other native orchids, to his garden.

17. Slipper Orchids are a bit different. Darwin regards them as possessing 'rudimentary' forms of some structures other species had developed more fully. They have, for example, no rostellum, possess three stigmas and two anthers, and have dusty pollen grains, not pollinia: *Orchids*, pp. 226–32.

18. Darwin, *Orchids*, p. 12.

19. As Darwin noted, 'the young flowers, which have their pollinia in the best state for removal, cannot possibly be fertilised; they must remain in a virgin condition until they are a little older and the column has moved away from the labellum' (*Orchids*, p. 121).

20. Darwin, *Orchids*, p. 118.

21. Ibid., p. 119.

22. Ibid., p. 230.

23. Ibid., p. 231.

24. Ibid., p. 49.

25. Ibid., pp. 281–2.

26. Ibid., p. 52.

27. Ibid., p. 58.

28. Endersby, pp. 90–1.

29. Sprengel called these deceptive orchids *Scheinsaftblumen*, 'sham-nectar-producers'. Darwin called the notion a 'gigantic . . . imposture' (Darwin, *Orchids*, p. 37). He was wrong.

30. Edens-Meier and Bernhardt, p. 7; Johnson, D., p. 24.

31. Vereecken and Francisco, pp. 52–67; Baguette et al., pp. 1632–52; Schiff, pp. 91–109.

32. Vereecken and Francisco, pp. 47–52.

33. Endersby, pp. 102–3.

34. Zhang et al., p. 380. Research by Li et al. suggests the earliest appearance of certain genomes in the orchid family emerged nearer 90 million years ago (pp. 7–8).

35. Arditti, 'Orchids', p. 74. Claessens and Kleynen, 'Investigations', offers a full exploration of the Bee's pollination strategy (pp. 62–76). They note that insects play an exceedingly small role. See also Claessens and Kleynen, *Flower*, pp. 330–8.

36. This movement is more pronounced and faster on the bright yellow pollinia and caudicles of the Early Purple Orchid.

37. Darwin, *Orchids*, p. 79.

Chapter 9: Buried Horizons

1. Darwin, *Orchids*, p. 277.

2. Ibid., pp. 277–8.

3. See Yam and Arditti, pp. 8–15.

4. Ibid., pp. 17–19; Anghelescu et al., p. 526.

5. Gerard (1597), p. 176.

6. Darwin, *Orchids*, p. 125.

7. Sheldrake, pp. 140–55.

8. Swarts and Dixon, *Conservation Methods*, pp. 31, 37, 56.

9. Darwin, Orchids, p. 103.

10. See Bersweden (pp. 307–12), Dunn (pp. 312–21), and Krulwich (n.p.) for accounts of the Ghost's history.

11. Claessens and Kleynen, *Flower*, p. 116; note that the Ghost has recently been found (unexpectedly) colonising pinewoods in continental Europe.

12. Swarts and Dixon, *Conservation Methods*, p. 56.

13. Kauth et al., p. 375.

14. Yeung, 'A Perspective on Orchid Seed', pp. 8–12, Rasmussen, pp. 120–3, and Withner, pp. 135–44 offer more information on protocorms and their mycorrhizal infection.

15. The operation of orchid mycorrhizal fungi and orchid germin-
    ation – pathways of infection, fungus species, structural events in
    germination and what fungi actually do – is complex and remains
    inadequately understood. See for example Rasmussen, pp. 98–112
    and 142–73, for an overview.
16. Ibid., pp. 314, 276, 238.
17. Hillel, pp. 27–8.
18. WWF, p. 48, states that a quarter of all life on earth resides
    underground.
19. Darwin, *Origin*, p. 369.
20. Harrap and Harrap, p. 137.

Chapter 10: Growing Dust

1. Yam and Arditti, 'History of Orchid Propagation', p. 20. Prior
   to Ramsbottom and Charlesworth many growers and scientists
   had attempted to germinate seed with varying degrees of failure
   (Yam and Arditti, 'History of Orchid Propagation', pp. 2–17).
2. Ibid., pp. 22–5.
3. See, for example, Rasmussen; Kauth et al.; Nabieva; Kitsaki et
   al.

Chapter 11: A Noseful of Lizard

1. Gerard (1597), p. 160.
2. Parkinson, p. 1348.
3. Gerard (1597), p. 161. Fifty years later it seems their medical usage
   had been reinstated: Parkinson (p. 1349) references the Flemish
   physician Dodoens, who 'saith that the rootes of these Orchies
   [*sic*] are better than any of the other, for the purposes aforesaid'.
   Those 'purposes' being 'to procure lust' (p. 1346).
4. Darwin, *Orchids*, p. 25.
5. Harrap and Harrap, p. 360.
6. In recent years the Lizard has been making a natural resurgence,
   probably due to climate change (see Appendix).

Chapter 12: Burnt-tips and Longhorns

1. Harrap and Harrap, p. 352.
2. Johnson's 1633 edition of *The Herball*, p. 207. A year later, Johnson, *Mercurius*, specifically records the species in England but gives it a different name – 'Litle [*sic*] purple-flowered Dogges-stones' (p. 54). He also notes that the species is present '*in montosis pratis*' ('in mountain pastures'). In this volume Johnson forgoes mention of the Austrian Dog stone.
3. Johnson, ed. Gilmour, pp. 1–4.
4. Harrap and Harrap, p. 352.
5. Ravetz and Turkington, p. 61.
6. See, for example, DEFRA, 'Biodiversity 2020: A Strategy'. By its own admission, most biodiversity indicators, such as the status of threatened habitats, status of priority species (abundance and distribution), breeding birds on farmland and in woodland, fish size in the marine environment and more 'deteriorated' up until 2021; however, this publication also includes some green ticks next to, for example, figures for increases in numbers of cattle and sheep breeds, non-governmental expenditure on biodiversity in the UK and the amount of data it has made available online, pp. 9–12.

Chapter 13: Environmental Outlaws

1. These, by the way, are random coordinates.
2. Black, n.p.
3. £600 million spent a year on treatment and prevention of illegal drugs; cost to society of illegal drugs is calculated as over £19 billion (Black, n.p). While the cost of drug use to individuals and to society is high – and receives significant media and police attention – in comparison 9 million people die prematurely every year from the effects of air pollution, mostly associated with the burning of fossil fuels; these deaths have a far higher personal, economic and social impact but comparatively insignificant media or police attention (Wallace-Wells, p. 103).

4. Wellsmith, pp. 10–12; Wildlife and Countryside Link, pp. 2–25, 30–41.
5. Crown Prosecution Service, 'Wildlife Crime Priorities', n.p.
6. Wildlife and Countryside Link, p. 34, 'Domestic crime relating to wild plants is not recorded and therefore no data is available. This makes it impossible to assess the true scale or nature of this type of crime.'
7. Stroh et al., p. 24.
8. In 2015 a 400-plus-page report compiled by the Law Commission reached the same conclusions: 'legislation governing the control, exploitation, welfare and conservation of wild animals and plants in England and Wales has become unnecessarily complex and inconsistent' (Law Commission, p. 2), and penalties for transgression require review to become 'more serious sanctions' (ibid., p. 4). In 2016 the then Parliamentary Under-Secretary of State at DEFRA, Thérèse Coffey, rejected the report's recommendations. (Coffey, n.p.).
9. Reid et al., pp. 128–35.
10. Tidman.
11. Shrubsole, p. 21. He also notes that 50 per cent of England is owned by less than 1 per cent of the population (p. 88).
12. Ibid., p. 97.
13. Ibid., p. 11. Monbiot, *Regenesis*, notes that 'Farming is the world's greatest cause of habitat destruction, the greatest cause of the global loss of wildlife, and the greatest cause of the global extinction crisis' (p. 90). It is also important to recognise that destructive farming methods emerged to supply continually growing demand.
14. Shrubsole, p. 233. Shrubsole also notes that of the 5.4 per cent of England made up of domestic homes and gardens, only about 70 per cent of those homes are owner-occupied.
15. Ibid., pp. 109–36.
16. Harrap and Harrap, p. 351.
17. Barkham.
18. BBC News, 'Millions of Homes'. See also Barton et al., pp. 5–6.

Chapter 14: Kentish Monkeys

1. Gerard (1597), p. 157.
2. Wilks, p. 51.
3. Ibid., pp. 51–3.
4. Dunn, p. 56.
5. Gerard (1597), p. 164.
6. Dunn, p. 164; Bersweden, p. 151.
7. Darwin, *Orchids*, p. 71; Claessens and Kleynen, *Flower*, pp. 277–80.

Chapter 15: Sainsbury's Slippers

1. Dunn, p. 126, notes that Linnaeus, who created the genus name *Cypripedium* (derived from *Cypris*, a synonym for the goddess Aphrodite, and *pedilon*, Greek for 'slipper'), got it slightly incorrect: it should be *Cypripedilum*.
2. Smith and Sowerby, vol. 1, plate 1, accompanying text.
3. Ibid.
4. Smith and Sowerby (1869 edn), vol. 9, p. 136.
5. Some doubts remain around whether this 'wild' plant may in fact have been 'of garden origin' (Harrap and Harrap, pp. 25–6). For more detailed accounts of this plant's history and rediscovery see Harrap and Harrap, pp. 30–1; Dunn, pp. 125–37.
6. Jenkinson, *Orchids of Hampshire*, pp. 12–13.
7. *Journal of the Kew Guild* (1989), p. 835.
8. Ibid. (1983), p. 227.
9. Ibid. (1989), p. 835.
10. Stewart, 'The Sainsbury Orchid Conservation Project', p. 41.
11. Stewart, *Orchids at Kew*, p. 129.
12. Stewart, 'The Sainsbury Orchid Conservation Project', p. 41.
13. *Journal of the Kew Guild* (1989), p. 835.
14. Stewart, 'The Sainsbury Orchid Conservation Project', pp. 41–2; Stewart, *Orchids at Kew*, p. 129.
15. Stewart, *Orchids at Kew*, p. 129. The Nature Conservancy Council ceased to operate in 1991 but, rebranded, English Nature took over the premises.

16. *Journal of the Kew Guild* (1989): the introduction of orchids at Wakehurst was 'to learn about conservation of our native terrestria [*sic*] orchids', p. 824.

17. Claessens and Kleynen, *Flower*, p. 122.

18. Stewart, 'The Sainsbury Orchid Conservation Project', p. 20.

19. Corkhill and Ramsay, n.p. Members of the Kew team also travelled to Germany to consult the successful German propagator Werner Frosch (Ramsay and Stewart, p. 178).

20. Stewart, *Orchids at Kew*, p. 123.

21. Harrap and Harrap, p. 213.

22. *Hardy Orchid Society Newsletter* 5 (1997), p. 13.

23. *Journal of the Kew Guild* (1997), p. 162.

24. Wheeler et al., pp. 142–45; Dunn, pp. 175–82; *Hardy Orchid Society Newsletter* 17 (2000), p. 12, notes that the few Fen Orchids planted back in the wild all died. Since then the charity Plantlife has had considerable success in reintroducing the Fen in Suffolk, Cambridgeshire, South Wales and the Norfolk Broads (Plantlife, 'Fen Orchid').

25. Paul Redshaw (personal communication).

26. Ramsay and Stewart, pp. 178–9.

27. *Hardy Orchid Society Newsletter* 7 (1998), p. 10.

28. Ibid., 11 (1999), p. 9.

29. Rasmussen, p. 232.

30. *Journal of the Kew Guild* (2001), p. 51.

31. Swainson, n.p.

32. Ibid.

33. Harrap and Harrap, p. 365.

34. Trudgill, pp. 17–22.

Chapter 16: The Sawfly Enigma

1. Fleming, pp. 190–1.

2. Brooke, *Wild Orchids*, p. 18.

3. McClintock, pp. 287–9.

4. Dunn, pp. 153–6 and Bersweden, pp. 133–6.

5. Brooke, *Wild Orchids*, p. 93.

6. Dunn, p. 155.

7. Brooke, *Wild Orchids*, p. 9.
8. Lousley, p. 360.
9. Dunn, p. 20.
10. Ibid., pp. 20–3.
11. Fowles, 'Interview', pp. 192–3.
12. Summerhayes, p. 1.
13. Brooke, *Wild Orchids*, pp. 15–17.
14. Summerhayes, p. 319.
15. Brooke, *Wild Orchids*, p. 18.
16. Harrap and Harrap, p. 382; and Cole and Waller, p. 216.

Chapter 17: Strimmers and Strategies

1. Last year that bank got strimmed.

Chapter 18: Samphire Hoe

1. Baguette et al., pp. 1644–5.
2. Samphirehoe.com
3. Harrap and Harrap, p. 412; Dunn, p. 30.

Chapter 19: Bog Angels

1. Parkinson, p. 505.
2. Koopowitz, p. 9.
3. A phrase used several times by the prime minister during and after the first waves of Covid-19 in Britain. It was epitomised in a policy paper (HM Treasury, 'Build Back Better').
4. HM Government, *Green Industrial Revolution*, p. 3.
5. Ibid., p. 5.
6. These various likely scenarios are discussed in Wallace-Wells ('Part II: Elements of Chaos'), pp. 39–140. As he notes, 'the threat from climate change is more total than from the bomb. It is also more pervasive' (p. 226).

7. In the early 2000s SOCP at Kew attempted to propagate Bog Orchids from their bulbils (at that time unsuccessfully): *Hardy Orchid Society Newsletter* 17 (2000), pp. 12–13. I don't know of anyone who has tried to grow their seeds *in vitro*.
8. Meteorological Office, 'Hottest Years'. This data has already been surpassed by more record-breaking temperatures in 2022.
9. Marsh, pp. 8–9.
10. Endersby, pp. 237–8.
11. Wraith and Pickering, pp. 15–16.

Appendix: Britain and Ireland's Wild Orchids

1. In the Appendix (and throughout this book) I sometimes refer to 'Britain' and 'British orchids' when these orchids may also be found in parts of Ireland and the United Kingdom. This is for ease, not to offend.
2. 'If a taxon is listed as being LC, it is important to emphasise that this does not imply that it is of no conservation concern, but rather that, in terms of extinction risk, it is not threatened' (Stroh et al., p. 9).
3. Despite my requests, various institutions and societies did not provide data from which more recent losses could be calculated.
4. Harrap and Harrap, p. 41.
5. Darwin, *Orchids*, pp. 125–7.
6. Jersáková et al., p. 2254.
7. Harrap and Harrap, p. 66.
8. Cole and Waller provide an overview of these hybrids and sub-species, pp. 180–213, 261–9.
9. Ibid., p. 30.
10. Summerhayes, p. 237.
11. Plantlife, 'Fen Orchids', n.p.
12. Harrap and Harrap, p. 266.
13. Marhold et al., pp. 159–60.
14. Darwin, *Orchids*, pp. 97–102.
15. Cole and Waller, p. 134.
16. Harrap and Harrap, p. 105.

SELECT BIBLIOGRAPHY

Adams, L. (1943), *Orchids*, L. Sherman Adams

Aldred, J. (2016, 9 September), 'Britain's Dormice Have Declined by Third Since 2000, Report Shows', *Guardian*, https://www.theguardian.com/environment/2016/sep/09/britains-dormice-have-declined-by-a-third-since-2000-report-shows [accessed 14 May 2022]

Allen, L., R. Reeve, A. Nousek-McGregor, J. Villacampa and R. MacLeod (2019), 'Are Orchid Bees Useful Indicators of the Impact of Human Disturbance?', *Ecological Indicators* 103, 745–55

Anghelescu, N., A. Bygrave, M. Georgescu, S. Petra and F. Toma (2020), 'A History of Orchids: A History of Discovery, Lust and Wealth', *Scientific Papers, Series B, Horticulture* 64:1, 519–30

Anon. (1862), review of C. Darwin, *On the Various Contrivances by which British and Foreign Orchids are Fertilized by Insects, and the Good Effects of Intercrossing*, in *Annals and Magazine of Natural History* 10 (3rd series), 384–88

Arditti, J. (1966), 'Orchids', *Scientific American* 212:4, 70–81

—— (2022), 'Orchids: Discoverers and Firsts', *South African Journal of Botany* 146, 815–48

Baguette, M., J. Bertrand, V. Stevens and B. Schatz (2020), 'Why are There so Many Bee-Orchid Species? Adaptive Radiation by Intra-Specific Competition for Mnesic Pollinators', *Biological Reviews* 95, 1630–63

Baldrian, P. (2019), 'The Known and the Unknown in Soil Microbial Ecology', *FEMS Microbiology Ecology* 95, 1–9

Bardgett, R. (2016), *Earth Matters: How Soil Underlies Civilization*, Oxford University Press

Barkham, P. (2015, 18 May), '97% of Britain's Wildflower Meadows Have Gone: Here's Why it Matters', *Guardian*, https://www. theguardian.com/commentisfree/2015/may/18/losing-97-percent-britain-wildflower-meadows-matters-butterfly [accessed 23 May 2022]

Barton, C., L. Booth and W. Wilson (2022, 4 February), 'Tackling the Under-Supply of Housing in England', House of Commons research briefing

BBC News (2019, 8 January), 'England "Needs Millions of Homes to Solve Housing Crisis"', https://www.bbc.co.uk/news/uk-england-46788530 [accessed 25 June 2022]

Bernhardt, P. and R. Edens-Meier (2014), 'Darwin's Orchids (1862, 1877): Origins, Development, and Impact', in R. Edens-Meier and P. Bernhardt (eds), *Darwin's Orchids: Then and Now*, University of Chicago Press, 3–20

Bersweden, L. (2017), *The Orchid Hunter: A Young Botanist's Search for Happiness*, Short Books

BirdGuides (2019, 13 April), 'Council Accidentally Bulldozes Roadside Nature Reserve', https://www.birdguides.com/news/council-accidentally-bulldozes-roadside-nature-reserve/ [accessed 29 July 2022]

Black, C. (2020, 17 September), 'Review of Drugs: Summary', Home Office independent report, https://www.gov.uk/government/publications/review-of-drugs-phase-one-report/review-of-drugs-summary [accessed 29 July 2022]

Blackstone, J. (1737), *Fasciculus Plantarum Circa Harefield Sponte Nascentium*, Henry Woodfall

Botanical Society of Britain and Ireland Maps, https://bsbi.org/maps [accessed 3 January 2023]

Boyle, F. (1901, 2015 reprint), *The Woodlands Orchids, Described and Illustrated, with Stories of Orchid-Collecting*, Jefferson Publications

Briggs, H. (2020, 17 December), '"World's Ugliest Orchid" Tops List of New Discoveries', BBC News, https://www.bbc.co.uk/news/science-environment-55339987 [accessed 25 June 2022]

Brooke, J. (1948–50, reprint 1981), *The Orchid Trilogy*, Penguin
—— (1951), *The Wild Orchids of Britain*, Bodley Head

Brown, R. (1831), *Observations on the Organs and Mode of Fecundation in Orchideae and Asclepiadeae*, Richard Taylor

Brundrett, M. and L. Tedersoo (2018), 'Evolutionary History of Mycorrhizal Symbioses and Global Host Plant Diversity', *New Phytologist* 220:4, 1108–15

Buglife (2022, 5 May), '"Bugs Matter" Survey Finds that UK Flying Insects Have Declined by Nearly 60% in Less than 20 Years', https://www.buglife.org.uk/news/bugs-matter-survey-finds-that-uk-flying-insects-have-declined-by-nearly-60-in-less-than-20-years/ [accessed 27 July 2022]

Carrington, D. (2019, 23 January), 'UK Has Biggest Fossil Fuel Subsidies in the EU, Finds Commission' *Guardian*, https://www.theguardian.com/environment/2019/jan/23/uk-has-biggest-fossil-fuel-subsidies-in-the-eu-finds-commission [accessed 27 July 2022]

—— (2019, 3 October), 'Populations of UK's Most Important Wildlife Have Plummeted Since 1970', *Guardian*, https://www.theguardian.com/environment/2019/oct/03/populations-of-uks-most-important-wildlife-have-plummeted-since-1970 [accessed 24 June 2022]

—— (2022, 6 January), '"Ghost" Orchid that Grows in the Dark Among New Plant Finds', *Guardian*, https://www.theguardian.com/environment/2022/jan/06/ghost-orchid-that-grows-in-the-dark-among-new-plant-finds [accessed 25 June 2022]

Catton, R. (2012, 29 November), 'Bulldozers Destroy Thousands of Wild Orchids in Wolds', *York Press*, https://www.yorkpress.co.uk/news/10079101.bulldozers-destroy-thousands-of-wild-orchids-in-wolds/ [accessed 29 July 2022]

Champion, T. (2014), 'People in Cities: The Numbers', Foresight, Government Office for Science: Future of Cities working paper, https://assets.publishing.service.gov.uk/government/uploads/system/uploads/attachment_data/file/321814/14-802-people-in-cities-numbers.pdf [accessed 25 June 2022]

Claessens, J. and J. Kleynen (2002), 'Investigations on the autogamy in *Ophrys apifera* Hudson', *Jahresberichte des Naturwissenschaftlichen Vereins Wuppertal* 55, 62–77

—— (2011), *The Flower of the European Orchid: Form and Function*, Claessens and Kleynen

Clements, R. (2008), 'The Common Kestrel Population in Britain', *British Birds* 101:5, 228–34

Cocker, M. (2018), *Our Place: Can We Save Britain's Wildlife Before it is Too Late?*, Jonathan Cape

Coffey, T. (2016, 22 November), Letter to Rt Hon. Lord Justice Bean, https://s3-eu-west-2.amazonaws.com/lawcom-prod-storage-11jsxou24uy7q/uploads/2015/04/20161122-Minister-Coffey-to-Rt-Hon-Lord-Justice-Bean.pdf [accessed 4 November 2022]

Cole, S. and M. Waller (2020), *Britain's Orchids: A Field Guide to the Orchids of Great Britain and Ireland*, Princeton University Press

Corkhill, P. and M. Ramsay (n.d.), 'Foreign Travel to Sweden', unpublished report for English Nature

Cozzolino, S. and A. Widmer (2005), 'Orchid Diversity: An Evolutionary Consequence of Deception?', *Trends in Ecology and Evolution* 20:9, 487–94

Cribb, P. (2014), obituary of Lady Sainsbury, *Orchid Research Newsletter* 64, 2–3

Crown Prosecution Service (2022, 8 January), 'Wildlife Offences: Legal Guidance', https://www.cps.gov.uk/legal-guidance/wildlife-offences [accessed 30 July 2022]

Culpeper, N. (1653; 1816 edn), *The British Herbal and Family Physician, to Which is Added a Dispensatory for the Use of Private Families*, M. Garlick

Curtis, W. (1777–98), *Flora Londinensis, or, Plates and Descriptions of Such Plants as Grow Wild in the Environs of London*, vols 1–7, William Curtis

Darwin, C. (1859; 1998 edn), *On the Origin of Species by Means of Natural Selection, or The Preservation of Favoured Races in the Struggle for Life*, Wordsworth

—— (1862, 1877, 1885), *The Various Contrivances by Which Orchids are Fertilised by Insects*, Elibron Classics (facsimile of 1885 edn published by John Murray)

—— (1958), *The Autobiography of Charles Darwin 1809–1882*, ed. N. Barlow, Collins

—— (1993), *The Correspondence of Charles Darwin*, 8: *1860*, Cambridge University Press

—— (1994), *The Correspondence of Charles Darwin*, 9: *1861*, Cambridge University Press

—— (2002), *The Correspondence of Charles Darwin*, 13: *1865 – Supplement to the Correspondence 1822–1864*, Cambridge University Press

Davies, S. (2019, May), 'Bringing Beavers Back: A Brief History', Scottish Wildlife Trust, https://scottishwildlifetrust.org.uk/2019/05/bringing-beavers-back/ [accessed 24 June 2022]

Davis, J. (2018, 21 July), 'Rewilding Distilled', Rewilding Earth, https://rewilding.org/rewilding-distilled/ [accessed 24 June 2022]

—— (2020, 26 September), 'UK Has "Led the World" in Destroying the Natural Environment', Natural History Museum, https://www.nhm.ac.uk/discover/news/2020/september/uk-has-led-the-world-in-destroying-the-natural-environment.html [accessed 28 July 2022]

DEFRA (Department for Environment, Food and Rural Affairs) (n.d.), 'Environmental Management: Wildlife and Habitat Conservation, Detailed Information', https://www.gov.uk/topic/environmental-management/wildlife-habitat-conservation [accessed 16 June 2022]

—— (2011), 'Biodiversity 2020: A Strategy for England's Wildlife and Ecosystem Services', policy paper, https://assets.publishing.service.gov.uk/government/uploads/system/uploads/attachment_data/file/69446/pb13583-biodiversity-strategy-2020-111111.pdf [accessed 24 June 2022]

—— (2016, 21 March), 'British Food and Farming at a Glance', https://assets.publishing.service.gov.uk/government/uploads/system/uploads/attachment_data/file/515048/food-farming-stats-release-07apr16.pdf [accessed 20 June 2022]

—— (2020), *UK Biodiversity Indicators 2020*, https://data.jncc.gov.uk/data/048f7e78-a2c6-4982-91c3-e496f063bf2b/UKBI-2020-A.pdf [accessed 20 November 2022]

—— (2021, 21 March), 'Consultation Launched on Environmental Principles: Five Legally Binding Principles Will Guide Future Policymaking to Protect the Environment', press release, https://www.gov.uk/government/news/consultation-launched-on-environmental-principles [accessed 16 June 2022]

—— (2021), *UK Biodiversity Indicators 2021*, https://hub.jncc.gov.uk/assets/31925413-3fa4-4382-962f-6ac2c168d542#ukbi2021-summary-booklet.pdf [accessed 28 July 2022]

—— (2022, 12 May), 'Draft Environmental Principles Policy Statement', policy paper, https://www.gov.uk/government/publications/environmental-principles-policy-statement/draft-environmental-principles-policy-statement#the-5-environmental-principles [accessed 27 July 2022]

Department for Levelling Up, Housing and Communities (2021, 5 November), '£624 Million of Loan Funding to Support Thousands of New Homes and Improve Vital Infrastructure', press release, https://www.gov.uk/government/news/624-million-of-loan-funding-to-support-thousands-of-new-homes-and-improve-vital-infrastructure [accessed 14 May 2022]

Desmond, A. and J. Moore (1991), *Darwin*, W. W. Norton

Dressler, R. (1961), 'The Structure of the Orchid Flower', *Missouri Botanical Garden Bulletin* 49, 60–69

Duffy, K., M. Waud, B. Schatz, T. Petanidou and H. Jacquemyn (2019), 'Latitudinal Variation in Mycorrhizal Diversity Associated with a European Orchid', *Journal of Biogeography* 46:5, 968–80

Dunn, J. (2018; 2019 edn), *Orchid Summer: In Search of the Wildest Flowers of the British Isles*, Bloomsbury

Easton, M. (2018, 3 January), 'The Illusion of a Concrete Britain', BBC News, https://www.bbc.co.uk/news/uk-42554635 [accessed 24 June 2022]

'Economic History: Farm-gardening and Market Gardening' (2004), in P. Croot (ed.), *A History of the County of Middlesex*, xii: *Chelsea, Victoria County History*, 150–155, British History Online, http://www.british-history.ac.uk/vch/middx/vol12/ [accessed 16 June 2022]

Edens-Meier, R. and P. Bernhardt (eds) (2014), *Darwin's Orchids: Then and Now*, University of Chicago Press

Endersby, J. (2016), *Orchid: A Cultural History*, University of Chicago Press and Kew Publishing

Environmental Performance Index (2020), 'Biodiversity Habitat Index', https://epi.yale.edu/epi-results/2020/component/bhv [accessed 30 July 2022]

Evans, R. (2019, 17 April), 'Half of England is Owned by Less Than 1% of the Population', *Guardian*, https://www.theguardian.com/money/2019/apr/17/who-owns-england-thousand-secret-landowners-author [accessed 20 June 2022]

Fay, M. and I. Taylor (2015), 'Cypripedium Calceolus', *Curtis's Botanical Magazine* 32:1, 24–32

Fay, M., M. Feustel, C. Newland and G. Gebauer (2018), 'Inferring the Mycorrhizal Status of Introduced Plants of *Cypripedium calceolus* (Orchidaceae) in Northern England Using Stable Isotope Analysis', *Botanical Journal of the Linnean Society* 186:4, 587–90

Fleming, I. (1963, reprinted 2004), *On Her Majesty's Secret Service*, Penguin Classics

Foley, M. (1992), 'The Current Distribution and Abundance of *Orchis Ustulata* L. (Orchidaceae) in the British Isles: An Updated Summary', *Watsonia* 19, 121–26

Fowles, J. (1979; reissued 2010), *The Tree*, Ecco

—— (1999), 'Interview with James R. Baker' in *Conversations with John Fowles*, ed. D. Vipond, University Press of Mississippi

Gaskett, A. (2014), 'Color and Sexual Deception in Orchids: Progress toward Understanding the Functions and Pollinator Perception of Floral Color', in R. Edens-Meier and P. Bernhardt (eds), *Darwin's Orchids: Then and Now*, University of Chicago Press, 291–310

Gerard, J. (1597), *The Herball, or Generall Historie of Plantes*, John Norton

—— (1633 edn), *The Herball, or Generall Historie of Plantes, Gathered by John Gerarde of London, Master in Chirurgerie, Very Much Enlarged and Amended by Thomas Johnson Citizen and Apothecarye*, Adam Norton and Richard Whitakers

Godfery, M. (1933), *Monograph and Iconograph of Native British Orchidaceae*, Cambridge University Press

Goulson, D. (2021), *Silent Earth: Averting the Insect Apocalypse*, Jonathan Cape

Government Office for Science (2021, 28 June), 'Trend Deck 2021: Urbanisation', https://www.gov.uk/government/publications/trend-deck-2021-urbanisation/trend-deck-2021-urbanisation [accessed 25 June 2022]

Gow, D. (2020), *Bringing Back the Beaver: The Story of One Man's Quest to Rewild Britain's Waterways*, Chelsea Green

Griffiths, M. (2002), *Orchids: From the Archives of the Royal Horticultural Society*, Scriptum Editions

Haggar, J. (2012), 'Propagating Orchids from Seed from an Amateur's Perspective', *Journal of the Hardy Orchid Society* 9:2 (64), 59–66

Hansen, E. (2000), *Orchid Fever: A Horticultural Tale of Love, Lust and Lunacy*, Methuen

Harbron, N. and L. Harbron (2012), '*Orchis Mascula*: Observations and Questions', *Journal of the Hardy Orchid Society* 9:1 (63), 19–21

Hardy Orchid Society Newsletters (1996–2003)

Harrap, A. and S. Harrap (2nd edn, 2009), *Orchids of Britain and Ireland: A Field and Site Guide*, A & C Black

Hayhow, D., M. Eaton, A. Stanbury, F. Burns, W. Kirby, N. Bailey, B. Beckmann, J. Bedford, P. Boersch-Supan, F. Coomber, E. Dennis, S. Dolman, E. Dunn, J. Hall, C. Harrower, J. Hatfield, J. Hawley, K. Haysom, J. Hughes, D. Johns, F. Mathews, A. McQuatters-Gollop, D. Noble, C. Outhwaite, J. Pearce-Higgins, O. Pescott, G. Powney, N. Symes (2019), *State of Nature 2019*, State of Nature partnership

Hillel, D. (1992), *Out of the Earth: Civilization and the Life of the Soil*, Aurum Press

Hinsley, A., H. de Boer, M. Fay, S. Gale, L. Gardiner, R. Gunasekara, P. Kumar, S. Masters, D. Metusala, D. Roberts, S. Veldman, S. Wong and J. Phelps (2018), 'A Review of the Trade in Orchids and its Implications for Conservation', *Botanical Journal of the Linnean Society* 186:4, 435–55, https://doi.org/10.1093/botlinnean/box083 [accessed 25 June 2022]

HM Government (1981), Wildlife and Countryside Act 1981, https://www.legislation.gov.uk/ukpga/1981/69/contents [accessed 14 May 2022]

—— (2006), Natural Environment and Rural Communities Act 2006, https://www.legislation.gov.uk/ukpga/2006/16/introduction/enacted [accessed 25 June 2022]

—— (2007), The Conservation (Natural Habitats, &c.) Amendment (Scotland) Regulations 2007, https://www.legislation.gov.uk/ssi/2007/80/regulation/29 [accessed 14 June 2022]

—— (2008), Climate Change Act 2008, https://www.legislation.gov.uk/ukpga/2008/27/contents [accessed 27 July 2022]

—— (2017), The Town and Country Planning (Environment Impact Assessment) Regulations 2017, https://www.legislation.gov.uk/uksi/2017/571/contents/made [accessed 29 July 2022]

—— (2020, November), *The Ten Point Plan for a Green Industrial*

Revolution: Building Back Better, Supporting Green Jobs, and Accelerating Our Path to Net Zero, https://assets.publishing.service.gov.uk/government/uploads/system/uploads/attachment_data/file/936567/10_POINT_PLAN_BOOKLET.pdf

—— (2021), Environment Act 2021, https://www.legislation.gov.uk/ukpga/2021/30/contents/enacted [accessed July 27 2022]

—— (2021, 10 November), 'World-Leading Environment Act Becomes Law', press release, https://www.gov.uk/government/news/world-leading-environment-act-becomes-law [accessed 27 July 2022]

—— (2022), Homes England: Land Hub, https://experience.arcgis.com/experience/c6f225e5589f498fa458f1f7a8bbbcb2 [accessed 22 January 2023]

HM Treasury (2020, 12 March), 'Budget 2020', policy paper, https://www.gov.uk/government/publications/budget-2020-documents/budget-2020 [accessed 30 July 2022]

—— (2021, 3 March), 'Build Back Better: Our Plan For Growth', policy paper, https://www.gov.uk/government/publications/build-back-better-our-plan-for-growth/build-back-better-our-plan-for-growth-html [accessed 27 July 2022]

Hooker, W. (1805), *The Paradisus Londinensis: or Coloured Figures of Plants Cultivated in the Vicinity of the Metropolis*, 2 vols, W. Hooker

Horton, H. (2019, 14 April), 'Brexit No Deal Planning has Destroyed Thousands of Britain's Rarest Orchids, it Emerges', *Sunday Telegraph*, https://www.telegraph.co.uk/news/2019/04/14/brexit-no-deal-planning-has-destroyed-thousands-britains-rarest/ [accessed 29 July 2022]

Hughes, O. (2018), 'Orchid–Mycorrhiza Relationships, Propagation of Terrestrial and Epiphytic Orchids from Seed', PhD thesis, Manchester Metropolitan University

IEA (International Energy Agency) (2021), *World Energy Investment 2021*, IEA, https://www.iea.org/reports/world-energy-investment-2021 [accessed 25 June 2022]

IUCN (International Union for Conservation of Nature and Natural Resources) (2001), *IUCN Red List Categories and Criteria, Version 3.1*, IUCN

IUCN UK (2021), 'Peatland Progamme', https://www.iucn-uk-peatlandprogramme.org/about-peatlands/uk-peatlands [accessed July 24 2022]

Jacquemyn, H., R. Brys, M. Waud, P. Busschaert and B. Lievens (2015), 'Mycorrhizal Networks and Coexistence in Species-Rich Orchid Communities, *New Phytologist* 206:3, 1127–34

Jacquemyn, H., M. Waud, V. Merckx, B. Lievens and R. Brys (2015), 'Mycorrhizal Diversity, Seed Germination and Long-term Changes in Population Size Across Nine Populations of the Terrestrial Orchid *Neottia ovata*', *Molecular Ecology* 24:13, 3269–80

Jenkinson, M. (1991), *Wild Orchids of Dorset*, Orchid Sundries Ltd

—— (1995), *Wild Orchids of Hampshire and the Isle of Wight*, Orchid Sundries Ltd

Jepson, P. and C. Blythe (2020), *Rewilding: The Radical New Science of Ecological Recovery*, Icon Books

Jersáková, J., J. Minasiewicz and M. Selosse (2022), 'Biological Flora of Britain and Ireland: *Neottia nidus-avis*', *Journal of Ecology* 301, 2246–63

Johnson, D. (2019), *Wild Orchids of Kent*, Kent Field Club

Johnson, T. (1849), *Mercurius botanicus: Sive plantarum gratiâ suscepti itineris, anno MDCXXXIV*, ed. T. Ralph, Pamplin

—— (1972), *Botanical Journeys in Kent and Hampstead: A Facsimile Reprint with Introduction and Translation of his Iter Plantarum 1629, Descriptio Itineris Plantarum 1632*, ed. J. Gilmour, Hunt Botanical Library

Joint Nature Conservancy Council (2019), UK BAP Priority List Species, https://jncc.gov.uk/our-work/uk-bap-priority-species/ [accessed 30 October 2022]

Journal of the Hardy Orchid Society (2004–22)

Journal of the Kew Guild (1984–2017)

Kauth, P., D. Dutra, T. Johnson, S. Stewart, M. Kane and W. Vendrame (2008), 'Techniques and Applications of *In Vitro* Orchid Seed Germination', in J. Teixeira da Silva (ed.), *Floriculture, Ornamental and Plant Biotechnology,* v, Global Science Books, 375–91

Kitsaki, C., S. Zygouraki, M. Ziobora and S. Kintzios (2004), '*In Vitro* Germination, Protocorm Formation and Plantlet Development of Mature Versus Immature Seeds from Several *Ophrys* Species (Orchidaceae)', *Plant Cell Reports* 23, 284–90

Koopowitz, H. (2001), *Orchids and Their Conservation*, Batsford

Krulwich, R. (2016, 21 July), 'The Rarest Plant in Britain Makes a Ghostly Appearance', *National Geographic*, https://www.national-geographic.com/science/article/search-for-rare-british-ghost-orchid [accessed 24 June 2022]

Kuehn, R., H. Pedersen and P. Cribb (2019), *Field Guide to the Orchids of Europe and the Mediterranean*, Kew Publishing

Lavelle, P. (2012), 'Soil as a Habitat', in D. Wall et al. (eds), *Soil Ecology and Ecosystem Services*, Oxford University Press, 7–27

Law, C. (1967), 'The Growth of Urban Population in England and Wales, 1801–1911', *Transactions of the Institute of British Geographers* 41, 125–43

Law Commission (9 November 2015), 'Wildlife Law: Volume 1: Report', Law Com No. 362, https://s3-eu-west-2.amazonaws.com/lawcom-prod-storage-11jsxou24uy7q/uploads/2015/11/lc362_wildlife_vol-1.pdf [accessed 4 November 2022]

Lee, Y., E. Yeung, N. Lee and M. Chung (2006), 'Embryo Development in the Lady's Slipper Orchid, *Paphiopedilum delenatii*, with Emphasis on the Ultrastructure of the Suspensor', *Annals of Botany* 98:6, 1311–19

Li, Y., Z. Li, A. Schuiteman, M. Chase, J. Li, W. Huang, A. Hidayat, S. Wu, X. Jin (2019), 'Phylogenomics of Orchidaceae based on plastid and mitochondrial genomes', *Molecular Phylogenetics and Evolution* 139, 1–11

Linden, J. (1891–97), *Lindenia: Iconography of Orchids*, L. Linden

Lindley, J. (1838), *Sertum Orchidaceum: A Wreath of the Most Beautiful Orchidaceous Flowers*, J. Ridgway

—— (1852–1855), *Folio Orchidacea: An Enumeration of the Known Species of Orchids*, J. Matthews

Lousley, J. (1976), *Flora of Surrey*, David & Charles

McClintock, D. (1977), 'J. E. Lousley and Plants Alien in the British Isles', *Watsonia* 11, 287–90

McCormick, M. and H Jacquemyn (2014), 'What Constrains the Distribution of Orchid Populations?', *New Phytologist* 202, 392–400

Macdonald, B. (2020), *Rebirding: Restoring Britain's Wildlife*, Pelagic

Malmgren, S. and H. Nystrom (n.d.), 'Orchid Propagation', http://lidaforsgarden.com/orchids/engelsk.htm [accessed 14 May 2022]

Manning, P. (2012), 'The Impact of Nitrogen Enrichment on Ecosystems and Their Services', in D. Wall et al. (eds), *Soil Ecology and Ecosystem Services,* Oxford University Press, 256–69

Marhold, K., I. Jongepierová, A. Krahulcová and J. Kučera (2005), 'Morphological and Karyological Differentiation of *Gymnadenia densiflora* and *G. conopsea* in the Czech Republic and Slovakia', *Preslia* 77:2, 159–76

Marsh, L. (2022), 'BSBI New Year Plant Hunt 2022', *BSBI News* 150, 8–9

Massy, C. (2017), *Call of the Reed Warbler: A New Agriculture, a New Earth*, Chelsea Green

Masters, S. (2020, 29 February), 'Orchids are Woven through Chinese Culture. What Happens if they Vanish?', *Guardian*, https://www.theguardian.com/environment/2020/feb/29/orchids-are-woven-through-chinese-culture-what-happens-if-they-vanish [accessed 24 June 2022]

Meteorological Office (2019, 31 July), 'Top Ten UK's Hottest Years All Since 2002', Met Office, https://www.metoffice.gov.uk/about-us/press-office/news/weather-and-climate/2019/state-of-the-uk-climate-2018 [accessed 27 July 2022]

MHCLG (Ministry of Housing, Communities and Local Government) (2020), 'Land Use in England, 2018', fact sheet, https://assets.publishing.service.gov.uk/government/uploads/system/uploads/attachment_data/file/900974/Land_Use_in_England_2018__Fact_Sheet.pdf [accessed 24 June 2022]

—— (2021), National Planning Policy Framework, https://assets.publishing.service.gov.uk/government/uploads/system/uploads/attachment_data/file/1005759/NPPF_July_2021.pdf [accessed 24 June 2022]

Millican, A. (1891), *Travels and Adventures of an Orchid Hunter: An Account of Camp and Canoe Life in Colombia, While Collecting Orchids in the Northern Andes*, Cassell

Monbiot, G. (2014), *Feral: Rewilding the Land, Sea and Human Life*, Penguin

—— (2022), *Regenesis: Feeding the World Without Devouring the Planet*, Allen Lane

Morelle, R. (2010, 28 June), 'Conservationists Warn of Hay Meadow

Decline', BBC News, https://www.bbc.co.uk/news/10381309 [accessed 30 July 2022]

Nabieva, A. (2021), 'Asymbiotic Seed Germination and *In Vitro* Seedling Development of *Orchis Militaris*, an Endangered Orchid in Siberia', *Journal of Genetic Engineering and Biotechnology* 19:122

Natural England (2022a), 'Protected Plants, Fungi and Lichens: Advice for Making Planning Decisions', guidance, https://www.gov.uk/guidance/protected-plants-fungi-and-lichens-advice-for-making-planning-decisions [accessed 29 July 2022]

—— (2022b), 'Prepare a Planning Proposal to Avoid Harm or Disturbance to Protected Species', guidance, https://www.gov.uk/guidance/prepare-a-planning-proposal-to-avoid-harm-or-disturbance-to-protected-species [accessed 29 July 2022]

Nong, S. (1998), *The Divine Farmer's Materia Medica*, trans. Yang Shou-zhong, Blue Poppy Press

Norris, K., S. Buckland, R. Green, H. Roy and P. Stephens (2016), *Review of UK Biodiversity Indicators that Provide Status and Trends for Species*, report for DEFRA, http://nora.nerc.ac.uk/515118/1/N515118CR.pdf [accessed 25 June 2022]

O'Brien, J. (1911) *Orchids*, T. C. & E. C. Jack

Office for National Statistics (2021, July), 'Construction Development: Improvements to Regional and Sub-Sector Level Estimates, UK: July 2021', https://www.ons.gov.uk/businessindustryandtrade/constructionindustry/articles/constructiondevelopmentimprovementstoregionalandsubsectorlevelestimatesukjuly2021/2021-07-20 [accessed 16 June 2022]

Office of the Deputy Prime Minister (2005, 16 August), 'Biodiversity and Geological Conservation – Statutory Obligations and Their Impact Within the Planning System', government circular, https://assets.publishing.service.gov.uk/government/uploads/system/uploads/attachment_data/file/7692/147570.pdf [accessed 24 June 2022]

Oost, K. and M. Bakker (2012), 'Soil Productivity and Erosion', in D. Wall et al. (eds), *Soil Ecology and Ecosystem Services*, Oxford University Press, 301–14

Parkinson, J. (1640), *Theatrum Botanicum: The Theater of Plantes, or an Universall and Compleate Herball*, Tho. Cotes

Pearn, J. (2012), 'The Doctrine of Signatures, Materia Medica of Orchids, and the Contributions of Doctor-Orchidologists', *Vesalius* 18:2, 99–106

Pei, S. and Z. Yang (2018), 'Orchids and Its Uses in Chinese Medicine and Health Care Products', *Medical Research and Innovations* 2:1, 1–3

People's Trust for Endangered Species (2019), *Annual Review 2019*, https://ptes.org/wp-content/uploads/2020/07/Annual-review-for-web.pdf [accessed 25 June 2022]

Pesticide Action Network UK (2018, March), 'The Hidden Rise of UK Pesticide Use: Fact-Checking an Industry Claim', https://issuu.com/pan-uk/docs/the_hidden_rise_of_uk_pesticide_use?e=28041656/59634015 [accessed 24 June 2022]

Petrovan, S. and B. Schmidt (2016), 'Volunteer Conservation Action Data Reveals Large-Scale and Long-Term Negative Population Trends of a Widespread Amphibian, the Common Toad (*Bufo bufo*)', *PLOS One* 11(10): e0161943, https://doi.org/10.1371/journal.pone.0161943 [accessed 25 June 2022]

Plantlife (2019, 5 July), 'News: Plantlife Research Shows the Value – and Vulnerability – of Britain's Last Remaining Meadows', https://www.plantlife.org.uk/uk/about-us/news/plantlife-research-shows-the-value-and-vulnerability-of-britains-last-remaining-meadows [accessed 30 July 2022]

—— (2019, 10 September), 'News: Could Fen Orchid finally be Back from the Brink?' https://www.plantlife.org.uk/wales/about-plantlife-cymru/news-press/fen-orchid-back-from-the-brink [accessed 4 November 2022].

Powney, G., C. Carvell, M. Edwards, R. Morris, H. Roy, B. Woodcock and N. Isaac (2019, 26 March), 'Widespread Losses of Pollinating Insects in Britain', *Nature Communications* 10, 1018, https://doi.org/10.1038/s41467-019-08974-9 [accessed 25 June 2022]

Pridgeon, A. (2014), *Orchid Research Newsletter* 64

Rackham, O. (1986), *The History of the Countryside: The Classic History of Britain's Landscape, Flora and Fauna*, Weidenfeld and Nicolson

Ramsay, M. and J. Stewart (1998), 'Re-establishment of the Lady's Slipper Orchid (*Cypripedium calceolus* L.) in Britain', *Botanical Journal of the Linnean Society* 126, 173–81

Rasmussen, H. (1995), *Terrestrial Orchids: From Seed to Mycotrophic Plant*, Cambridge University Press

Rasmussen, H. and D. Whigham (1993), 'Seed Ecology of Dust Seeds in Situ: A New Study Technique and Its Application in Terrestrial Orchids', *American Journal of Botany* 80:12, 1374–78

Raum, S. (2020), 'Land-use Legacies of Twentieth-Century Forestry in the UK: A Perspective', *Landscape Ecology* 35, 2713–22

Ravetz, A. and R. Turkington (1995), *The Place of Home: English Domestic Environments 1914–2000*, Taylor & Francis

Ray, J. (1693) *Historia Plantarum, Tomus Secundus*, Smith and Walford
—— (1848), *The Correspondence of John Ray,* ed. E. Lancaster, Ray Society

—— (2011), *John Ray's Cambridge Catalogue (1690)*, ed. and trans. P. Oswald and C. Preston, Ray Society

Reichenbach, H. (1858–1900), *Xenia Orchidacea*, F. A. Brockhaus

Reid, C., K. Hornigold, E. McHenry, C. Nichols, M. Townsend, K. Lewthwaite, M. Elliot, R. Pullinger, A. Hotchkiss, E. Gilmartin, I. White, H. Chesshire, L. Whittle, J. Garforth, R. Gosling, T. Reed and M. Hugi (2021), *State of the UK's Woods and Trees 2021*, Woodland Trust

Reinikka, M. (1972), *A History of the Orchid*, University of Miami Press

Rewilding Britain (2022), 'Why We Need Rewilding', https://rewildingbritain.org.uk/explore-rewilding/what-is-rewilding/why-we-need-rewilding [accessed 25 June 2022]

Richardson, R. (2017), *Britain's Wild Flowers: A Treasury of Traditions, Superstitions, Remedies and Literature*, National Trust

Riley, J. (2014, 26 February), 'England's Lost World: 421 Species – Including Mammals, Birds and Plants – Have Become Extinct Over the Past 200 Years', *Daily Mail*, https://www.dailymail.co.uk/sciencetech/article-2568524/Englands-lost-world-421-species-including-mammals-birds-butterflies-extinct-past-200-years.html [accessed 30 October 2022]

Royal Botanic Gardens Kew (n.d.), 'Genetic Studies of *Cypripedium calceolus* (The Lady's Slipper Orchid)', https://www.kew.org/science/our-science/projects/genetic-studies-cypripedium-calceolus [accessed 16 June 2022]

Samphirehoe.com (2022), https://www.samphirehoe.com/ [accessed 20 August 2022]

Scardarella, L. (2021, 25 July), 'Rare Orchids Mown Down but Council Learns Lesson', *Wokingham Today*, https://wokingham.today/rare-orchids-mown-down-but-council-learns-lesson/ [accessed 16 June 2022]

Schiff, J. (2018), *Rare and Exotic Orchids: Their Nature and Cultural Significance*, Springer

Seaton, P. and M. Ramsay (2005), *Growing Orchids From Seed*, Royal Botanic Gardens Kew

Seaton, P., P. Cribb, M. Ramsay, J. Haggar (2011), *Growing Hardy Orchids*, Royal Botanic Gardens Kew

Seyler, B., O. Gaoue, Y. Tang, D. Duffy and E. Aba (2020), 'Collapse of Orchid Populations Altered Traditional Knowledge and Cultural Valuation in Sichuan, China', *Anthropocene* 29, 1–12

Sheldrake, M. (2020), *Entangled Life: How Fungi Make our Worlds, Change our Minds, and Shape our Futures*, Bodley Head

Shrubsole, G. (2019), *Who Owns England? How We Lost Our Green and Pleasant Land and How to Take it Back*, William Collins

Smith, J. and J. Sowerby (1798–1814), *English Botany: or Coloured Figures of British Plants with their Essential Characters, Synonyms, and Places of Growth*, J. Davis

—— (1869, third edn), *English Botany: or Coloured Figures of British Plants*, vol. 9, J Syme (ed.), Robert Hardwicke

Stanbury, A., A. Brown, M. Eaton, N. Aebischer, S. Gillings, R. Hearn, D. Noble, D. Stroud and R. Gregory (2017), 'The Risk of Extinction for Birds in Great Britain', *British Birds* 110, 502–517, https://www.bto.org/our-science/publications/peer-reviewed-papers/risk-extinction-birds-great-britain [accessed 27 July 2022]

Stewart, J. (ed.) (1992), *Orchids at Kew*, HMSO

—— (1993), 'The Sainsbury Orchid Conservation Project: The First Ten Years', *Kew Magazine* 10, 38–43

Stringer, C. (2022, 14 June), 'Blooming Disgrace! Council Mows Down Rare Orchids on Roundabout Despite 11-year-old Girl's Plea to Let Them Grow to Support Bees', *Daily Mail*, https://www.dailymail.co.uk/news/article-10917017/Council-mows-rare-orchids-roundabout-despite-11-year-old-girls-plea.html [accessed 29 July 2022]

Stroh, P., S. Leach, T. August, K. Walker, D. Pearman, F. Rumsey, C. Harrower, M. Fay, J. Martin, T. Pankhurst, C. Preston and I. Tayor (2014), *A Vascular Plant Red List for England*, Botanical Society of Britain and Ireland

Summerhayes, V. (1951), *Wild Orchids of Britain*, Collins

Sumption, K. and J. Flowerdew (2008), 'The Ecological Effects of the Decline of Rabbits (*Oryctolagus cuniculus* L.) due to myxomatosis', *Mammal Review* 15:4, 151–86

Swainson, G. (2022, 21 January), 'Gait Barrows NNR Orchid Update (from Natural England)', https://www.arnsidesilverdaleaonb.org.uk/gait-barrows-nnr-orchid-update-from-natural-england/ [accessed 23 July 2022]

Swarts, N. and K. Dixon (2009), 'Terrestrial Orchid Conservation in the Age of Extinction', *Annals of Botany* 104:3, 543–56

—— (2017), *Conservation Methods for Terrestrial Orchids*, J. Ross

Tali, K., M. Foley and T. Kull (2004), 'Orchid ustulata L.' (Biological Flora of the British Isles No 232), *Journal of Ecology* 92:1, 174–84

Taylor, I., S. Leach, J. Martin, R. Jones, J. Woodman and I. Macdonald (2021), *Guidelines for the Selection of Biological SSSIs. Part 2: Detailed Guidelines for Habitats and Species Groups. Chapter 11 Vascular Plants*, Joint Nature Conservation Committee

Taylor, L. and D. Roberts (2011), 'Biological Flora of the British Isles: *Epipogium aphyllum* Sw.', *Journal of Ecology* 99, 878–90

Teoh, E. (2019), *Orchids as Aphrodisiac, Medicine or Food*, Springer

Tree, I. (2019), *Wilding: The Return of Nature to a British Farm*, Picador

Trimen, H. and W. Dyer (1869), *Flora of Middlesex: A Topological and Historical Account of the Plants Found in the County*, Robert Hardwicke

Trudgill, D. (2022), 'An Exploration of Orchid Records in the BSBI Database in Four Regions of the British Isles', *BSBI News* 149, 16–22

Turner, W. (1881), *The Names of Herbes A.D. 1548*, N. Truebner

UK Biodiversity Action Plan (2007), 'Species and Habitats Review: De-Listed Species (2007)', spreadsheet, https://hub.jncc.gov.uk/assets/bdd8ad64-c247-4b69-ab33-19c2e0d63736#UKBAP-

DelistedSpecies–2007.xls [accessed 30 July 2022]

UK Centre for Ecology and Hydrology (2021, 18 November), 'Latest Land Cover Map Provides Greater Detail About British Landscape', https://www.ceh.ac.uk/latest-land-cover-map-provides-greater-detail-about-british-landscape [accessed 30 July 2022]

United Nations (1987), Resolutions Adopted on the Reports of the Second Committee, 43/196: United Nations Conference on Environment and Development, https://digitallibrary.un.org/record/153025 [accessed 30 October 2022]

Van Waes, J. M. and P. Debergh (1986), 'In Vitro Germination of Some Western European Orchids', *Physiologia Plantarum* 67:2, 253–61

Vera, F. (2000), *Grazing Ecology and Forest History*, CABI

Vereecken, N. and A. Francisco (2014), '*Ophrys* Pollination: From Darwin to the Present Day', in R. Edens-Meier and P. Bernhardt (eds), *Darwin's Orchids: Then and Now*, University of Chicago Press, 47–70

Veyret, Y. (1974), 'Development of the Embryo and the Young Seedling Stages of Orchids', in C. Withner (ed.), *The Orchid: Scientific Studies*, Wiley, 223–65

Vidal, J. (2008, 21 October), 'UK's Ancient Woodland Being Lost "Faster than Amazon"', *Guardian*, https://www.theguardian.com/environment/2008/oct/21/forests-conservation [accessed 24 June 2022]

Waite, S. (ed.) (1998), *Botanical Journal of the Linnean Society* 126:1–2

Wall, D., R. Bardgett, V. Behan-Pelletier, J. Herrick, T. Hefin Jones, K. Ritz, J. Six, D. Strong and W. van der Putten (eds) (2012), *Soil Ecology and Ecosystem Services*, Oxford University Press

Wall, W. and D. Morgan (2019), *How to Grow Native Orchids in Gardens Large and Small: The Comprehensive Guide to Cultivating Local Species*, Green Books

Wallace-Wells, D. (2019), *The Uninhabitable Earth: A Story of the Future*, Penguin

Watson, W. (1891), 'New or Little-known Plants: Dendrobium Phalaenopsis', *Garden and Forest: A Journal of Horticulture, Landscape Art and Forestry* 4, 520–22

Wellsmith, M. (2011), 'Wildlife Crime: The Problems of Enforcement', *European Journal on Criminal Policy and Research* 17, 125–48, https://doi.org/10.1007/s10610-011-9140-4 [accessed 25 June 2022]

Wheeler, B., P. Lambley and J. Geeson (1998), '*Liparis Loeselii* (L.) Rich. in Eastern England: Constraints on Distribution and Population Development', *Botanical Journal of the Linnean Society* 126, 141–58

Wild Flower Society, 'Code of Conduct for the Conservation and Enjoyment of Wild Plants', https://bsbi.org/wp-content/uploads/dlm_uploads/Code_of_Conduct.pdf. [accessed 3 October 2022]

Wildlife and Countryside Link (2021), *Wildlife Crime in 2020: A Report on the Scale of Wildlife Crime in England and Wales*, https://www.nwcu.police.uk/wp-content/uploads/2021/11/WCL_Wildlife_Crime_Report_Nov_21.pdf [accessed 24 June 2022]

Wildlife Trusts (2020, 15 January), *What's the Damage? Why HS2 Will Cost Nature Too Much*, http://57ac0c4836212612790f-823d78da580330fdca9d1ec4339d74c1.r33.cf1.rackcdn.com/What's%20the%20damage%20-%20Full%20Report%20digital.pdf [accessed 30 July 2022]

Wilks, H. (1960), 'The Re-discovery of *Orchis Simia* Lam. in Kent', *The Transactions of the Kent Field Club* 1:2, 50–55

Willis, K. and C. Fry (2015), *Plants: From Roots to Riches*, John Murray

Wilson, A. (2017), *Charles Darwin: Victorian Mythmaker*, John Murray

Withner, C. (1974), 'Developments in Orchid Physiology', in C. Withner (ed.), *The Orchids: Scientific Studies*, Wiley, 129–168

—— (ed.) (1974), *The Orchids: Scientific Studies*, Wiley

Wood, H. (1883), *A Season Among the Wild Flowers*, W. S. Sonnenschein

Woolf, J. (2020), *Britain's Trees: A Treasury of Traditions, Superstitions, Remedies and Literature*, National Trust

Wraith, J. and C. Pickering (2019), 'A Continental Scale Analysis of Threats to Orchids', *Biological Conservation* 234, 7–17

Wurst, S., G. De Deyn and K. Orwin (2012), 'Soil Biodiversity and Functions', in D. Wall et al. (eds), *Soil Ecology and Ecosystem Services*, OUP, 28–44

WWF (2018), *Living Planet Report – 2018: Aiming Higher*, M. Grooten and R. Almond (eds), WWF

Yam, T. and J. Arditti (2009), 'History of Orchid Propagation: A Mirror of the History of Biotechnology', *Plant Biotechnology Report* 3, 1–56

Yam, T., J. Arditti and K. Cameron (2009), '"The Orchids Have Been a Splendid Sport": An Alternative Look at Charles Darwin's

Contribution to Orchid Biology', *American Journal of Botany* 96:12, 2128–54

Yam, T., H. Nair, C. Sin Hew and J. Arditti (2022), 'Orchid Seeds and their Germination: An Historical Account', in T. Kull and J. Arditti (eds), *Orchid Biology: Reviews and Perspectives VIII*, Springer, 387–504

Yeung, E. (2017), 'A Perspective on Orchid Seed and Protocorm Development', *Botanical Studies* 58:33

Yeung, E., J. Park and I. Harry (2018), 'Orchid Seed Germination and Micropropagation I: Background Information and Related Protocols', in Y. Lee and E. Yeung (eds), *Orchid Propagation: From Laboratories to Greenhouses – Methods and Protocols*, Springer Protocols Handbooks, 101–25

Yuan, S.-C., S. Lekawatana, T. Amore, F.-C. Chen, S.-W. Chin, D. Monge Vega and Y.-T. Wang (2021), 'The Global Orchid Market', in F.-C. Chen and S.-W. Chin (eds), *The Orchid Genome,* Springer Nature, 1–28

Zayed, Y. and P. Loft (2019, 25 June), 'Agriculture: Historical Statistics', Government Briefing Paper No. 03339, https://research-briefings.files.parliament.uk/documents/SN03339/SN03339.pdf [accessed 24 June 2022]

Zhang, G.-Q., K.-W. Liu, Z. Li et al. (2017), 'The *Apostasia* Genome and the Evolution of Orchids', *Nature* 549, 379–83, https://doi.org/10.1038/nature23897 [accessed 24 June 2022]

Bird's-nest Orchid, from J. Smith and J. Sowerby,
English Botany, vol. 1 (1790).

INDEX